Building Terminology

by the same author

Formwork and Concrete Practice

BUILDING TERMINOLOGY

An illustrated reference guide for
practitioners and students

Peter Brett

Heinemann Newnes

Heinemann Newnes
An imprint of Heinemann Professional Publishing Ltd
Halley Court, Jordan Hill, Oxford OX2 8EJ

OXFORD LONDON MELBOURNE AUCKLAND SINGAPORE
IBADAN NAIROBI GABORONE KINGSTON

First published 1989

British Library Cataloguing in Publication Data

Brett, Peter, 1950–
    Building terminology.
    1. Construction.
    I. Title
    624

ISBN 0 434 90176 8

Photoset by Wilmaset, Birkenhead, Wirral
Printed and bound in Great Britain by
Courier International Ltd, Tiptree, Essex

Contents

Preface

The purpose of this guide is to provide in an understandable, well-illustrated form, the exact definition of the main range of terms used daily in the building industry and, in addition, others of a more historical interest.

This invaluable and authoritative guide has been compiled to supplement the more specialist textbooks which deal with various specific building subject areas and to provide a single source of information which gives an overview of the building industry as a whole. As such it can be considered an essential text for practitioners and all students studying City and Guilds of London Institute building craft, advanced craft and supplementary studies courses – not only for the technology and associated subjects but also for assignment work and the common industrial studies component.

This guide will also be useful as a reference for students on building technician, architectural and other associated courses with a building element, both in the UK and overseas.

Although written for practitioners and students this guide will be of value to anyone connected with the building industry and members of the general public who may require information on a specific topic.

Peter Brett

How to use this guide

The terms defined are grouped together under a number of topic areas. To find the definition of a known term, turn to the appropriate section and look up the term, which will be found in its alphabetical position within the section.

Where the topic area is unknown the alphabetical index at the end of the guide can be used to find the page reference. On the page will be the word with its clear definition and reference to related words to see. In addition, illustrations are incorporated to aid understanding. Further cross references are given to other sections where appropriate.

The definitions of terms which are illustrated are shown within a tinted block. Where a term is included as part of another illustration the number in brackets at the end of the definition indicates the page on which it appears.

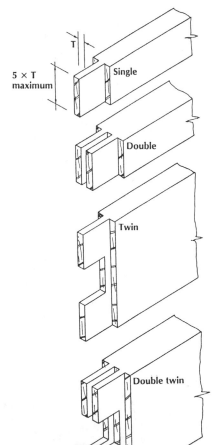

Example

Tenon   The end of a framing member that is reduced in thickness. To be inserted into a mortise, in another member to which it is to be joined. The thickness of a tenon should be approximately one-third of the thickness of the timber to be joined and its width not more than five times its thickness. *See also* Double, Twin, Haunch, Stub and Tusk tenons.

This term is illustrated adjacent to the entry.

Twin tenon   Two tenons arranged within the width of a piece of timber separated by a haunch, one following the other. Normally used for jointing the middle and bottom rails of doors to the stiles. See also Double tenon. (157)

This term is included in an illustration on page 157.

1
ARCHITECTURAL STYLE

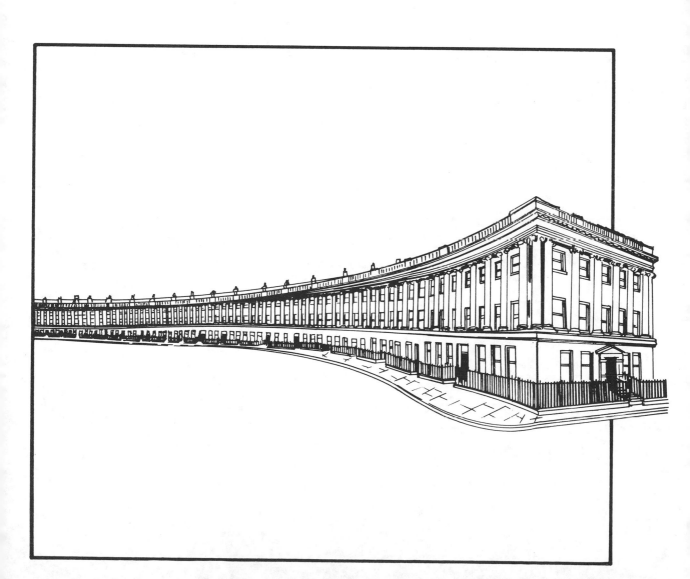

Abacus The slab at the top of a column capital. *See also* Orders. (25)

Abbey Originally the dwelling of the head of a monastery or convent. Later used to include the church and all buildings associated with a monastic establishment.

Abutment A pier from which an arch springs. Also the junction where building elements meet.

Acanthus A spiny, thick-leaved plant used to model the leaves for the capital of Corinthian columns. *See also* Orders.

Acanthus

Adam's style A classical style of architecture, furniture and interior decoration associated with Robert Adam in the eighteenth century. Laid out much of London, introduced stucco to England – main features include moulded plaster, spider's web fanlights, white fireplaces and fine pictures on ceilings. (27)

Adam's style

Aedicule An opening (door, window or shrine etc.) framed by columns or pilasters and topped by an entablature and pediment. Such an opening is said to be aediculated.

Aesthetics The beauty or otherwise of an object.

Aisle An area of a church flanking the nave on either side. Divided from the nave by arcades. (41)

Alcove Recess in a room set aside to hold a bed. Often closed off with doors or railed off with a balustrade.

Alley An aisle, a narrow street or passage in a built-up area, an enclosed garden walkway.

Alms house A dwelling provided for the shelter of poor persons.

Ambulatory A walkway or church aisle.

Amphitheatre An elliptical or circular building with rising tiers of seats. Originally used by the Romans for public spectacles.

Ancient light A window that has had continuous access to light for a substantial number of years. It is legally entitled to the continuance of that light, unobstructed by the construction of any new building, structure or addition.

Aedicule

Ancones The decorated brackets or consoles with support cornices above openings in classical buildings.

Anta A classical pilaster, the base and capital of which does not conform to the order of adjacent columns. These usually occur at the ends of the flanking walls of a portico. If there are columns between the flanking walls they are said to be in antis. *See also* Prostyle. (23)

Antechamber or anteroom A room which is used to provide access to a further room beyond.

Anticum The porch to a front door.

Apartment A single room or suite of rooms that are self-contained and form a complete dwelling.

Apse A semi-circular, arched or domed recess. *See also* Church.

Aqueduct An elevated trough or channel for the conveyance of water. *See also* Viaduct.

Araeostyle One of five forms of intercolumniation, having an arrangement of columns spaced four shaft diameters apart. *See also* Diastyle, Eustyle, Pycnostyle and Systyle.

Arcade A series of arches supported on columns or piers. *See also* Arch. (41)

Arch A curved structure bridging an opening and capable of supporting a load. Built of separate components each supporting one another by mutual pressure. Variously named according to shape.

Architect One who designs a building. *See also* heading under Documentation, administration and control.

Architecture A distinctive style of building. Also the science of building design.

Architrave The lowest division of the entablature resting directly on the supporting column capitals. Also used to describe the trim around doors and windows which covers the joint between wall and frame. (23, 43)

Arcuated A building which uses the arch principle as its form of construction, as opposed to the post and beam trabeated method.

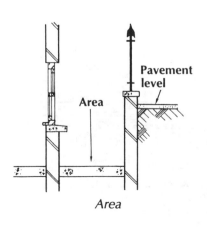

Area

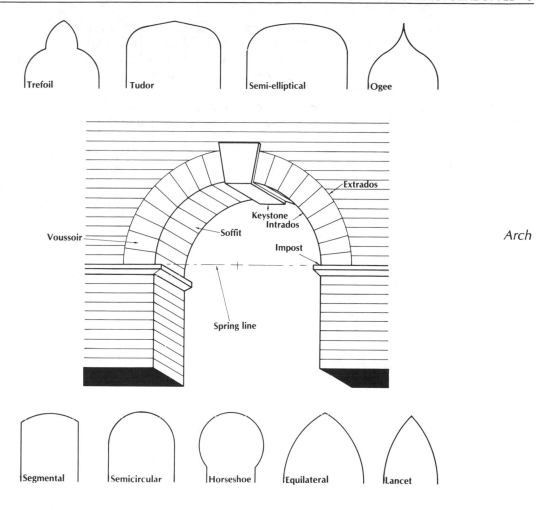

Arch

Area   The sunken space around a building to allow light into the basement storey and prevent damp penetration. Also used to mean the surface of a space or the superficial extent of a figure.

Arena   The central area of an amphitheatre where the contest or action took place.

Armoury   A storage place for weapons.

Arris   The sharp edge or corner where two surfaces join.

Arsenal   A public storage place for weapons.

Art Deco   The decorative style of the 1920s and 1930s which derives its name from a Paris exhibition of decorative and industrial arts in 1925. It is characterized by geometric design and bright metalic surfaces.

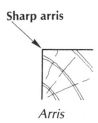

Arris

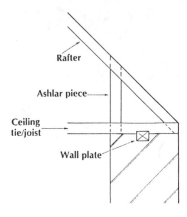

Ashlaring

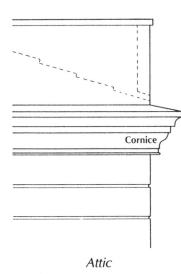

Attic

Art Nouveau A style of decorative art which reached its height between 1890 and 1910. It was a move away from the imitation of past style. Characterized by curved designs based on either natural or abstract forms.

Articulation The design of a building that gives clarity or distinction to the individual architectural members.

Arts and crafts A nineteenth century movement based on the respect for medieval craftsmanship. First developed by William Morris who set up cooperative workshops to produce furniture, textiles, pottery and wallpaper.

Ashlar Squared stone blocks laid in regular parallel courses.

Ashlaring or ashlar piece The short, vertical timbering between ceiling joists and rafters which cuts off the acute angle in a roof space.

Astragal A small, semicircular moulding or bead. *See also* Torus.

Astylar A classical facade without columns or pilasters.

Atrium An open central courtyard or hall within a building and surrounded on all sides by roofed areas.

Attic The storey or parapet wall of a classical building above the cornice, originally intended to hide the roof slope. May also be in the form of a balustrade.

Awning A covering to protect persons or parts of a building from the sun or rain.

Back to back   Victorian factory workers' homes terraced on three sides having windows only on the front. Soon labelled as slums they shared a communal water tap and earth toilet.

Bailey   The walls surrounding a keep, or a courtyard enclosed by fortified walls.

Balcony   A platform projecting from the wall of a building usually sited below windows or doors and protected by a balustrade.

Balistraria   A narrow slit (often cruciform) in medieval walls, through which bowmen fired arrows.

Baluster   A small decorative column supporting a coping or handrail and forming part of a balustrade. (8)

Balustrade   A structure that guards an open edge consisting of balusters and supporting a coping or handrail. *See also* heading under Building construction.

Band   A string course or projecting moulding running across the façade of a building. (27)

Balistraia

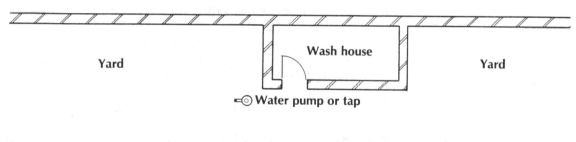

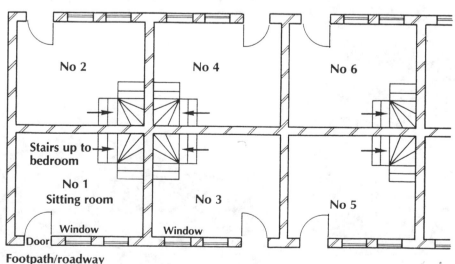

Back to back (early workers' back to back housing)

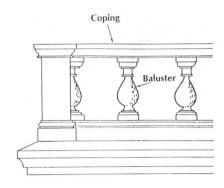

Balustrade

Banister   A baluster. A series of banisters forms a balustrade.

Barbican   A fortified gateway or tower forming the outer defence of a city or castle.

Bargeboard   The decoratively carved boards that are often fixed to the overhanging gable ends of roofs. *See also* heading under Building construction.

Barn   A covered farm building used for storage.

Baroque   Early period of Georgian architecture known as the age of elegance. Blenheim Palace is the major example of grand, highly decorative style. *See also* Rococo.

Barracks   A building used to house troops.

Barrel vault   A continuous vault, semicircular in section, also known as a tunnel vault.

Baroque

Internal fibrous plaster decoration

Blenheim Palace

Bartisan   A turret.

Base   The part of a column between its shaft and pedestal or pavement. *See also* Orders. (21, 43)

Basement   The first storey of a building partly or wholly below ground level.

Basilica   Originally a Roman hall of justice. Its rectangular plan shape consisting of nave, flanking aisles and apse was later used in church design.

Bastille   A fortified wall or tower often used as a prison.

Bastion   A tower at the corner of a fortified wall.

Battle of styles   The competition between classical style architecture and Gothic revival architecture during the early Victorian period, when both were battling for dominance.

Battlement   A notched or indented parapet wall used to protect troops. The raised parts are called cops or merlons and the gaps crenelles or embrasures. *See also* Castellated, Crenellated and Embattled.

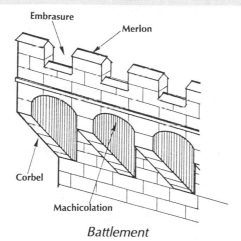

Battlement

Bay   The space enclosed between the main structural element of a building marked by crucks, trusses, buttresses, columns or pilasters etc.

Bay leaf garland   A classical enrichment on torus mouldings.

Bay stall   A seat in a bay window.

Bay window   A window projecting from the face of a building and forming a recess in the room. Named according to shape. A bay window that projects from an upper storey only, is called an oriel. *See also* heading under Building construction.

Bead   Any small, cylindrical or partly cylindrical moulding.

Bead and butt   A panel that finishes flush with its surrounding framing. Its vertical edges having a bead stuck on while the horizontal edges simply butt to the rails.

Beak head   A decoration to Norman mouldings consisting of beak animal heads.

Beehive huts   Bronze and iron age stone, dome-shape dwellings.

Belfry   The upper part of a tower that houses bells.

Bellcote   A turret or gable used to hang bells.

Belt   A band that projects from a façade.

Belvedere   A turret, lantern or room built at high level to provide a view. *See also* Gazebo.

Billet   A Norman ornamental moulding.

Bird's beak   An ogee or ovolo moulding, the outline of which resembles a bird's beak.

Black and white work   Timber-frame structures where the blacked timber work stands out in contrast to the white, limewashed wall infill. Associated with the Tudor period.

Blank   A false door or window. *See also* Blind.

Blind   Tracery or arcading which is purely decorative, being applied directly on a wall surface without any glazing or openings. Also used to describe a door or window opening that has been filled in.

Blind storey   The triforium of a church. It has no windows. Can also be applied to a tall parapet wall or false storey that conceals a roof.

Boast   To carve or shape a piece of timber or stone to its original shape prior to building it in, at which point its fine detail can be carved at leisure.

Bolection   A moulding that stands proud of its surrounding framework. (27)

Boss   An ornamental block used to cover the intersection of the ribs to a vault.

Boudoir   A withdrawing room used by the lady of the house.

Boultine   An ovolo or convex moulding.

Bow   A projecting part of a building having a segmental or semicircular plan. *See also* Bow window and Apse.

Bow window   A bay window having a segmental or semicircular plan shape. *See also* Bow.

Bowtell   An astragal moulding.

Box   A small room or dwelling.

Bracket   A projection from the face of a wall to carry a shelf, seat etc. or support a statue or ornament. May also be called a corbel or bragger.

Bracketing   The cradling or wooden framework of brackets which provides a fixing for the lathing or large plaster cornices.

Bragger   A bracket.

Brandering   Counter battening fixed to the underside of floor joists to provide good level fixing for ceiling laths or boards.

Breast   The part of a chimney that projects into a room.

Breastsummer   A large lintel spanning a wide opening (supporting brickwork or masonry) often over a shop or bay window; formally of timber now concrete or steel.

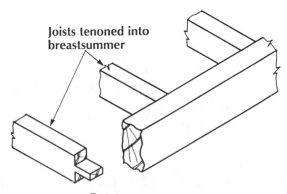

Joists tenoned into
breastsummer

Breastsummer

Broach Traditionally any spire but mainly applied to an octagonal spire that rises from a square tower without parapets. (53)

Brutalism A term used to describe the 1950s contemporary style which was characterized by the extensive use of unfinished concrete surfaces.

Building A structure having an external envelope that encloses space. *See also* heading under General.

Building

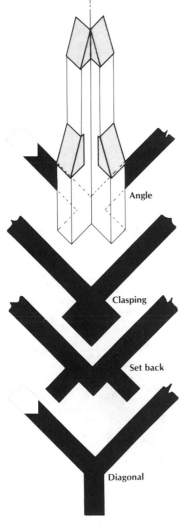

Angle

Clasping

Set back

Diagonal

Buttress

Bungalow A one-storey house. Originally a light, thatched structure occupied by Europeans in India.

Butment The same as abutment.

Buttery A storeroom for wine and provisions. Can also be applied to wine bars and other places where food and wine is served.

Buttress A short cross wall or a thickening of a wall forming a projection in order to provide additional strength and support. Variously named according to their positioning. For example, angle, clasping, set back and diagonal buttress. *See also* Flying buttresses.

Byzantine A style of architecture that was developed in the eastern Roman Empire at Byzantium (later Constantinople) in AD 330.

Cable A Romanesque moulding taking the form of a twisted strand of rope.

Caisson A coffer or cofferdam.

Canopy A covering hung over a bed, throne or altar, or any similar covering over any other object.

Cant An external angle of a building that is not a right angle e.g. a cant bay window. *See also* Canted.

Canted Any member that is splayed, bevelled or off square. A polyagonal column can be termed a canted or cant column.

Cantilever A beam, bracket or other member that is securely supported at one end only and carries a load at its free end.

Cap The uppermost part of any member e.g. a capital, cornice or coping etc.

Capital The uppermost part of a column or pilaster. Named according to their associated order. *See also* Cushion capital. (23, 43)

Caryatid A sculptured column in the form of a female figure.

Casino A public room or building used for dancing, music or gambling. Can also be applied to a small country house, summerhouse or decorative lodge.

Castellated Having battlements. *See also* Crenellated and Embattled.

Castle A fortified building used for defence purposes. *See also* Motte, Keep and Donjon.

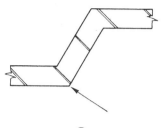

Cant

Castle

Catacomb   An underground room or place used for storing the dead.

Cathedra   A bishop's throne or chair.

Cathedral   The principle church or diocese in which the bishop's cathedra is located.

Causeway   A raised bank or dam forming a roadway across water or marshy ground.

Cavedium   An open court within a house.

Cavetto

Cavetto   A hollow or concave moulding, the profile of which forms a quandrant of a circle. *See also* Scotia.

Cell   A small room in which to sit, often a bedsitting room as part of a prison or monastery.

Cellar   The lowest storey of a building which is either partly or wholly underground. *See also* Basement.

Cemetery   A place for burying dead bodies.

Cenotaph   An empty tomb or monument that celebrates the memory of dead people lost or buried elsewhere.

Cesspool or pit   A hole in the ground used for the collection of human wastes.

Chair rail   A dado rail or moulding fixed around a room at chairback height to prevent damage to the wall surface.

Chamber   A room e.g. bed chamber, great chamber and royal chamber etc. *See also* Chamber storey.

Chamber storey   The storey of a building that contains the bed chambers.

Chambranle   A decorated border around doors and windows.

Chamfer

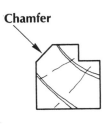

Chamfer

Chamfer   An arris or angle that is cut off at forty-five degrees. It is symmetrical. *See also* Splay.

Chancel   The eastern part of a church containing the choir stalls. Part of church from which those who officiate conduct services.

Chandry   A place used for storing candles.

Chapel   A room or small building used for worship.

Chapter house   A meeting hall. A building connected to a cathedral where the dean, prebendaries and canons met when carrying out business.

Charnel house   A building used to store old bones cleared from a churchyard, in order to reuse the ground for fresh graves.

Chequerwork   A decorative treatment of walls and floors in a chessboard pattern.

Chevron   A zig-zag decoration used in Norman buildings. *See also* Dancette.

Chinoiserie   A style of architecture, decoration and furniture having a Chinese influence. Common in the rococo.

Chippendale   A Georgian furniture designer/maker.

Choir   The part of a church which has stalls for singers.

Choir stalls   Elevated seats in the chancel of a church used by the singers.

Church   A building used for public Christian worship.

Cimborio   A raised lantern light over a roof or fenestrated cupola.

Cinquefoil   A tracery pattern consisting of five foils. *See also* Cusp.

Cippus   A low column or post often bearing an inscription. Used as milestones and monuments.

Circus   A terrace or group of houses arranged in the form of a circle. Also a circular building or one with semicircular ends used for games. *See also* Crescent.

Cistern   Storage tank or reservoir. *See also* heading under Services and finishes.

Citadel   A dominating fortress, situated in a fortified town or city and forming the final defence or retreat for the defending party.

Clasping buttress   A square on plan buttress located at the corner of a building. (12)

Chippendale chair

Classical architecture   Architecture using the classical principles of Greek and Roman buildings. Classical style is based on symmetry and the concept of orders, each part being mathematical in proportion and related to each other. *See also* Renaissance.

Clerestory or clear-story   The upper storey of a church containing windows to light the nave. Can be applied to any light in the upper part of a wall. (41)

Clink   A prison.

Cloister   A covered passageway or ambulatory surrounding an open quadrangle giving access to several buildings or apartments.

Closet   A small room connected to a bed chamber.

Clustered   Several members grouped together e.g. a clustered column or pier.

Cob wall   A wall constructed from layers of pressed mud, gravel and straw.

Cockle stair   A winding stair.

Coffer   A sunken ceiling panel. Also termed a caisson.

Cofferdam   A watertight structure used to dam rivers and sea etc. Enables foundations and walls to be built below water level. Also termed a caisson.

Coffin   A box to receive a dead body.

College   A building of further/higher education or part of a university. Traditionally consisting of one or more quadrangles around which the rooms were arranged.

Colonnade   A row of columns variously named according to their grouping or arrangement e.g. distyle, tetrastyle, hexastyle, octastyle, decastyle. *See also* Peristyle, Portico and Columniation.

Colonnette   A small column.

Colossal order   *See* Giant order.

Colosseum   A large Roman amphitheatre.

Columellae   Balusters.

Colonnade

Column A vertical structural member, traditionally cylindrical, consisting of a base, shaft and capital, supporting an entablature. They are the major distinguishing features of any particular style. *See also* Colonnade, Orders and Pilasters. (43)

Columniation An arrangement of columns or a building with columns. *See also* Intercolumniation.

Comitium A building used for assembly purposes.

Common house or room A room in a monastery, university or college etc. where a fire was always alight to provide a warm welcome.

Communion table A wooden table in churches used for bread and wine.

Compartment ceiling A panelled ceiling.

Composite order A mixture of Ionic and Corinthian orders used by Roman builders. *See also* Orders.

Compound A clustered member.

Compound arch A series of concentric arches.

Concamerate Meaning to arch over.

Conch A semidome of an apse or a niche.

Composite order

Confessio A recess in a wall used to display objects.

Confessional A recess in a wall, a box or a seat, used by a priest to hear confession.

Conservatory A building used to house or conserve plants, more grand than a greenhouse. The term is also used for buildings where music or the arts are practised.

Console An ornamented bracket or corbel. Also called ancones.

Convent A building used by nuns or other religious orders.

Corbel A block projecting from a wall face to support a beam, arch springing or parapet. *See also* Corbelling. (9)

Corbelling The use of corbels to support a feature above.

Corinthian order An order of architecture used by both Greek and Roman builders. Characterized by its ornate acanthus leaf column capitals.

Cornice The top projecting member of an entablature; the moulding at the top of an external wall or over openings to throw off rainwater; also the moulding used internally at the wall/ceiling junction. (6, 21, 23, 43)

Corridor A galley or passageway giving access to separate rooms.

Counterfort A buttress.

Court A walled open space before, behind or in the centre of a building; or an open space surrounded by buildings; or a place for administering law.

Cote A Saxon peasant's cottage built using crucks.

Cottage A small house in the country used by farm labourers or artisans.

Credence A shelf or table placed to the side of an altar for sacraments.

Crematorium A building or place used for burning dead bodies.

Crenellated Having battlements.

Corinthian order

Crepido An ornament of a cornice that projects from the surface.

Crescent A terrace of dwellings in a curved form. Called a circus when they form a complete circle.

Crescent

Crest A badge or emblem applied to a building e.g. family crest; or an ornamental finish to the top of a building element e.g. crusted ridge tiles.

Crinkle-crankle A wall having a serpentine plan shape.

Crockets Projecting carved leaf forms used for Gothic ornamentation.

Croft A small farm or its cottage. Also a crypt. *See also* Undercroft.

Crope The bunch of leaves that terminates a spire, also termed a finial.

Cross The symbol of the Christian religion.

Cross vaulting The intersection formed by two or more vaults.

Crossing The area of a church where the nave, transepts and chancel intercept. Often topped or crowned by a dome or tower.

Crow steps Gable walls extending past the roof line and finishing in a stepped fashion.

Crow steps

Crown   The highest point of a building.

Crown glass   Glass formed by blowing moulten glass into a balloon shape, spinning it to form a disc of glass known as a crown and finally cutting it up into pieces. The centre marked piece was considered substandard and sold off cheaply, although it is now considered a popular olde-worlde feature.

Cruciform   Having the shape of a cross.

Cruck frame houses   A Saxon method of housebuilding using crucks covered with wattle and daub and thatch.

Crucks   Curved split tree trunks forming the pointed, arch end frames of cruck frame houses.

Crypt   A vaulted room beneath a building either partly or completely below ground level. Also termed a croft or undercroft. (41)

Cullis   A channel, chase or groove. *See also* Portcullis.

Cupboard   Traditionally a board or shelf on which cups, silver plate and other items could be displayed. Now used to mean any piece of furniture that contains shelves for storage. *See also* Potboard.

Cupola   A small cup-like domed roof. Also a concave ceiling.

Curtain wall   The outer wall of a castle. *See also* heading under Building construction.

Curvilinear tracery   Decorated Gothic tracery.

Cushion capital   A Romanesque column capital taking the form of a cube, having its lower corners rounded off to fit the circular shaft.

Cusp or cusping   A point formed by the meeting of two curves or foils. These project into the glazed area of decorated tracery.

Cyma   A moulding used in classical architecture having an S-shape consisting of a convex and a concave curve. *See also* Cyma recta and Cyma reversa. (41)

Cyma recta   An S-shape moulding having its concave part uppermost. Also known as an ogee mould. *See also* Cyma.

Cyma reversa   The reverse of a cyma recta. An S-shape mould having its convex part uppermost. *See also* Cyma.

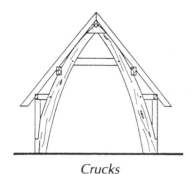

Crucks

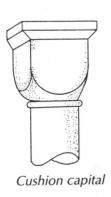

Cushion capital

Dado or die In classical architecture the solid block forming the pedestal or plinth, between the base and cornice. Also applied to the finishing of the lower part of interior walls from about waist height down to the floor. *See also* Dado rail under Building construction.

Dagger A motif resembling a dagger used in decorated tracery.

Dairy A building or room used to store milk and manufacture butter and cheese etc.

Dais Any raised platform. Originally used for the raised platform, or raised table at the end of a medieval hall where the family and guests sat.

Dancette A zig-zag or chevron decoration.

Dark Ages The Middle Ages.

Days The lights of a window.

Decastyle A portico or colonnade having ten frontal columns.

Decorated style The middle period of English, pointed Gothic style architecture from about AD 1200–1300. Characterized by lavish ornamentation. (32)

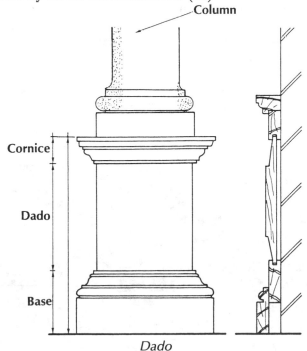

Dado

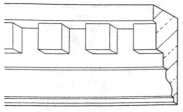

Dentils

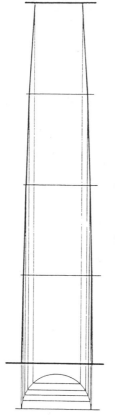

Diminuation of a column

Dentils The rectangular or cubic, teeth-like projections used as ornamentation to classical mouldings. *See also* Orders.

Diaperwork A decorative diamond pattern in face brickwork used in Gothic style architectures, created using lighter or darker coloured headers.

Diaphragm arch or vault An arch and its wall above, across the nave of a church to divide the timber roof into sections and thus acting as a fire break.

Diastyle One of five styles of intercolumniation, having an arrangement of columns spaced three shaft diameters apart. *See also* Araeostyle, Eustyle, Pycnostyle and Systyle.

Die A dado. *See also* Pedestal.

Diminuation of a column The tapering of a column with its height in order to create an elegant appearance. *See also* Entasis.

Diocletian window A Palladian-style window consisting of a semicircle divided by two vertical mullions to create three lights. Also termed a thermal window.

Dipteral A building with a double row of columns on each flank.

Distyle A portico or colonnade with two frontal columns.

Dodecastyle A portico or colonnade with twelve frontal columns.

Dogtooth An early English ornament consisting of raised, four-cornered, star-shaped leaves.

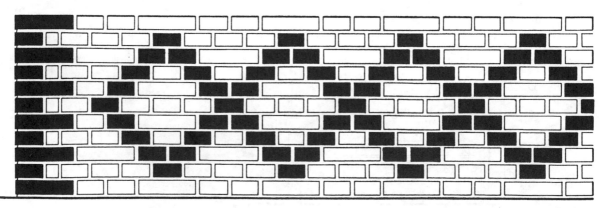

Diaperwork

Dome A vaulted roof; normally circular or polygonal in plan and semicircular, segmental or pointed in section. *See also* Cupola and Squinch.

Domus A Roman private home. *See also* Insula.

Donjon or dungeon The main tower of a castle. Also used for the small room or cell in the donjon used to house prisoners.

Doric order An order of architecture used by both Greek and Roman builders. Characterized by its fluted baseless columns.

Dormer A vertical window on a sloping roof surface having a roof of its own.

Dormitory A sleeping room with many beds.

Dovecot A building which houses doves or pigeons.

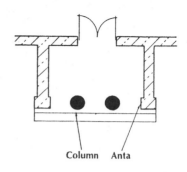

Plan of a distyle in antis portico

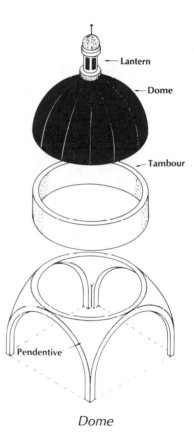

Dome

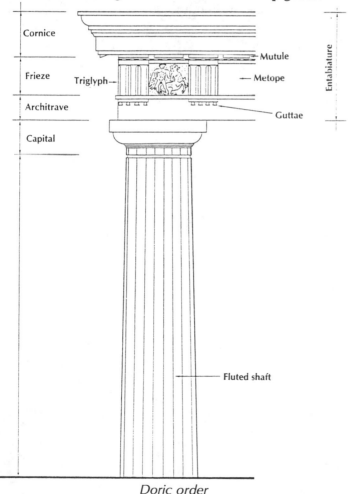

Doric order

Dragon beam   A beam of a traditional, timber-framed house that is set diagonally at the corner to support the corner post and floor joist where the building jetties on two adjacent sides.

Dragon piece   A diagonal tie across the wallplate at the corner of a hipped end roof for receiving the thrust of the hip rafter.

Drawbridge   A bridge capable of being raised and lowered.

Drawing room   A room originally called a solar or withdrawing room where the family or company could withdraw to privacy after eating.

Dressings   The mouldings that surround door and window openings, or any other feature.

Dripstone   A moulding projecting from the face of the wall over arches, doorways and windows, to throw off rainwater. Also called a label or hoodmould.

Dungeon   A donjon.

Early English The first period of English, pointed Gothic style architecture from about AD 1200–1300.

Echinus The ovolo moulding below the abacus of column capitals. Can be plain or covered with egg and dart enrichment.

Edge roll A rounded moulding.

Egg and dart An enrichment to ovolo mouldings consisting of alternate oval shapes and arrow or dart heads. Much used in the ionic order. *See also* Leaf and dart.

Elizabethan style The first period of architecture known as Renaissance, from 1558 to 1603. This saw the rediscovery of classical style which is based on symmetry and the concept of orders.

Embattled Provided with battlements, thus it is battlemented or crenellated.

Embossed Any sort of raised carved design or relief in brick, stone or timber.

Embrasure The gaps in a battlement; also the splayed reveals of doors and windows. (9)

Emplection Norman wall construction, consisting of two skins of smooth cut stone, filled with rubble.

Encarpus A frieze decoration consisting of festoons of fruit, flowers and leaves.

Encaustic tiles Glazed and decorated earthenware tiles, much used in Gothic and Victorian architecture.

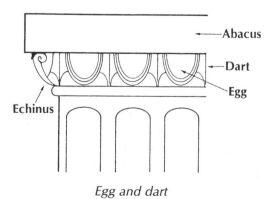

Egg and dart

Enfilade The alignment of internal doors connecting rooms so that long, through views are achieved when the doors are open. Introduced in Baroque-style architecture.

Entablature The upper part of an order that is the moulded beam or lintel spanning columns, consisting of architrave, frieze and cornice. (23, 43)

Entail Elaborate or finely sculptured ornaments.

Entasis The slight swelling or convex curvature of classical column shafts. Used to prevent the optical illusion of concavity which would be the result of straight shafts.

Entresol A mezzanine floor.

Eustyle One of five forms of intercolumniation, having an arrangement of columns spaced two and a quarter shaft diameters apart. *See also* Araeostyle, Diastyle, Pycnostyle and Systyle.

Ewery A scullery. A place where ewers (water jugs, dishes and pots) are washed.

Exchange A merchants' meeting place.

Exedra A recess with seats used for rest or conversation. Also an apse or large niche.

Extrados The exterior curved face of an arch. *See also* Intrados. (5)

Eye The centre of a building element or component.

Eyecatcher or folly A decorative building constructed in an English, landscaped park to terminate the view or create a picturesque effect. Often built as Gothic or classical sham ruins.

Façade   The exterior face or front of a building.

Fan light   The window over a door, often semicircular with radiating glazing bars, said to resemble a woman's open fan. A feature of Adam's style.

Fan vault   *See* Vault.

Fenestral   A window or component belonging to a window e.g. a shutter.

Fenestration   The arrangement of windows in the façade of a building.

Festoon   A sculptured wall decoration consisting of suspended swags of flowers, fruit, foliage and drapes.

Field or fielded   The raised centre section of any panel. *See also* Margin.

Fillet   The small, flat surface or band used between mouldings to separate them. (41)

Finial   The ornamental finishing which terminates canopies, gables, pediments and pinnacles.

Flags or flagged   Rectangular stone paving slabs.

Flank   The side or party wall of a building.

Fleur-de-lis   A lily-like ornamentation often used as a finial decoration.

Fleuron   Any carved floral decoration. Also termed floriated.

F

Fan light

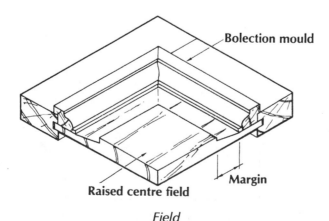

Field

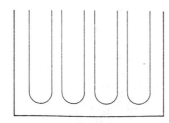

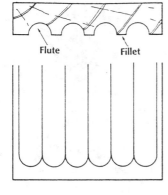

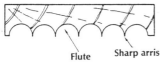

Fluting

Florentine arch   An arch where the intrados and extrados are not parallel.

Floriated   Any carved floral decoration. Also termed fleuron.

Flushwork   Use of knapped flint in panels with dressed stone to create decorative patterns.

Flutes or fluting   Shallow, concave, vertical grooves in the shafts of columns and pilasters etc. They may meet at sharp arrises or alternatively be separated by small fillets. *See also* Reed.

Flying buttress   A buttress in the form of a half arch which springs from an outer support or buttress. It transfers the thrust of upper vaults or roofs.

Foil   Part of a circle used in window and panel tracery. Cusps are formed where adjacent foils meet. The number of foils in a grouping is indicated by a prefix e.g. trefoil (three), quatrefoil (four), cinquefoil (five).

Foliage or foliated   Any carved leaf decoration.

Folly   A building positioned and constructed as an eye-catcher, often having no specific purpose.

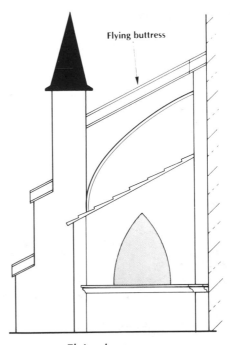

Flying buttress

Font   The vessel in a church containing the water which is used for baptism.

Fortress   A castle, military or other building constructed for defence purposes.

Forum   An open space surrounded by public buildings.

Foyer   The entrance hall or vestibule of a theatre or opera house.

Fresco   A decorative wall painting applied to the surface layer of plaster while it is still wet.

Fret   A geometric ornament of fillets meeting at right angles and repeated to form a band.

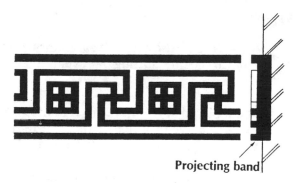

Projecting band

Fret

Frieze   The central division of the entablature located between the architrave and the cornice. It may be plain or decorated depending on the order. Also the decorated strip of an internal wall, directly below the cornice; the upper panels of a six-panel door are termed frieze panels; and the upper intermediate rail of a door can be termed a frieze rail. *See also* Encarpus. (23, 43)

Front   Any façade of a building, mainly used to refer to the elevation which contains the main entrance.

Frontispiece   The decorated main entrance of a building or its main façade.

Fronton   A pediment.

Functionalism   The belief that it is the main duty of an architect or designer to ensure that the building or structure functions well.

G

Gable or gavel   The end wall of a building which conforms to the slope of its abutting roof. *See also* Crow steps.

Gablet   A small gable, often ornamental.

Gadrooned   Vertical decoration of convex curves; exactly the reverse of fluted or fluting.

Gallery   A long balcony or mezzanine overlooking the main interior space of a building. Also a covered garden walkway and a building or room used to display works of art. *See also* Long gallery.

Galleting   The insertion of small pieces of stone, flint or brick into the wet mortar joints of a wall, mainly as a form of decoration.

Gaol   A prison.

Garderobe   A wardrobe. Also a medieval lavatory, often little more than a hole in the floor jutting out over a river, moat or street gutter.

Gargoyle   A spout placed at the eaves of a roof through which rainwater was discharged. Often took the form of an imaginary creature with water spouting from its throat.

Garland   A carved ornamental band of flowers and fruits.

Garret   The upper storey of a house either partly or wholly within the roof space.

Gauged arch   An arch constructed using precisely-cut voussoirs that radiate to a centre point.

Gazebo   A small summerhouse with a view, mainly situated in a park or garden. Can also be on the roof of a house where it should be termed a belvedere.

Georgian   The classical style of English architecture during the time of the first four Georges who ruled Great Britain from 1714 to 1830, encompassing Baroque, Palladian, Adam and Regency styles.

Gesso   A prepared plaster of Paris surface used as a background relief for painted designs.

Giant order   Any order in which the columns or pilasters rise through several storeys. Can also be termed colossal order.

Gargoyle

Gibbions An English woodcarver. The Royal Master Carver employed by Wren to decorate the choir stalls of St Paul's Cathedral.

Gibbs surround A door surround in the style of the architect James Gibb. It consists of large blocks of stone interrupting the architrave.

Gilding The application of gold leaf on a surface.

Gilloche A classical ornament to a band or convex moulding consisting of two or more intertwining bands forming a plait.

Gin palace Ornate public houses full of etched glass, advertisement mirrors and gas lights, dating from the 1830s. Built to encourage people away from their drab surroundings and also to combat the then recently introduced independent beer shops.

Golden section A proportion, thought to be divine by Renaissance architects. Defined as a line cut so that the minor section is proportional to the major section as the major section is proportional to the whole line. This is approximately a proportion of 5:8. It can be constructed geometrically using a right-angled triangle having one of its two shortest sides equal to the length of the line to be divided and the other half this length. A rectangle having sides of 5:8 can be considered as a golden rectangle. Golden rectangles can be found in the rooms and facades of many Renaissance buildings. *See also* Harmonic proportion.

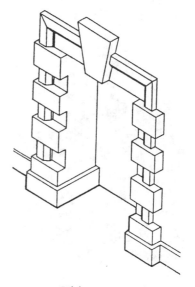

Gibbs surround

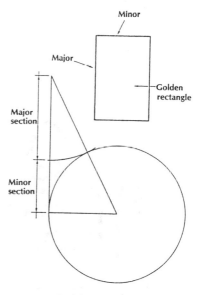

Golden section

Gilding

Gothic revival

Gothic architecture The style of architecture prevalent in Western Europe from 1200 to 1500. Characterized by pointed arches, flying buttresses, tracery, hammerbeam roofs, spires, vaulting and great churches. English Gothic is divided into three periods: Early English, Decorated and Perpendicular. The name Gothic was given to this period in the eighteenth century by followers of 'civilized' classical architecture in order to distinguish it from the 'barbaric' pointed style. *See also* Battle of styles and Gothick.

Gothic revival The revival of pointed architecture during the early Victorian period. *See also* Battle of styles and Gothick.

Gothick A term applied to the excessive use of Gothic ornamentation during the revival.

Graining The painting of a surface to imitate the grain of wood or the veining of marble.

Granary A building for storing grain or corn.

Grange A farm.

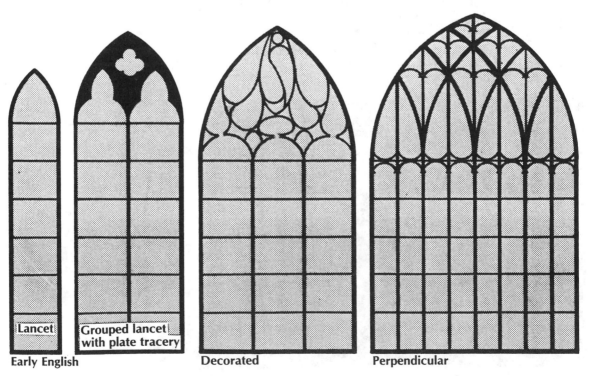

Early English — Lancet / Grouped lancet with plate tracery

Decorated

Perpendicular

Gothic architecture (windows)

Greek architecture A trabeated form of construction dating from about 600 BC onwards, and using the Doric, Ionic and Corinthian orders.

Greenhouse A glasshouse for propagating or growing plants that would not survive in the outside conditions. *See also* Conservatory.

Grees A staircase or steps.

Groin The sharp arris formed by the intersection of vaults.

Grotesque Intricate and fanciful ornamental decorations and consisting of medallions, figures and foliage. Used by the Romans as wall decoration. Found underground in 'grottos' hence their name.

Grotto An artificial cavern decorated with rock and shell-work, incorporating fountains and water cascades.

Grouped columns Two or more columns placed on the same pedestal.

Guttae Drop-like projecting ornaments placed under Doric cornice mutules and each triglyph on a Doric architrave. *See also* Orders. (23)

Gymnasium A place for athletic games and exercises. Also, in Ancient Greece, a place of teaching.

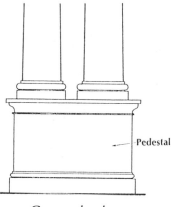

Grouped columns

Habitacle   A niche.

Half round   A semicircular or torus moulding.

Half timbered   A timber-framed building. The spaces between the timber members being filled with non-structural brickwork, wattle and daub or lath and plaster. *See also* Timber framing.

Hall   The main room of a dwelling in the Middle Ages.

Hammerbeam   A beam used in a Gothic roof in place of a tie beam but which does not extend across the entire span of the roof. Hammerbeam ends were often elaborately carved.

Harmonic proportion   A system of proportions used in classical architecture which relates architecture to music. If their measurements were of the ratios 1:2, 2:3, or 3:4 buildings or rooms were said to be harmonious. *See also* Golden section.

Helices   Volutes or spiral ornamentation.

Hermitage   A small dwelling or hut in a secluded location.

Herringbone   Bricks, stone or timber blocks that are arranged on a slant instead of flat. Often seen in paving, exterior walls and interior floors.

Hexastyle   A portico having six frontal columns.

Hippodrome   A Victorian music hall; originally a course for chariot races; also a place for staging equestrian events.

Hogging   A convex curved shape.

Hood mould   The dripstone or projecting moulding over window and door openings.

Horn   An Ionic volute. *See also* Orders.

Hospital   An institution for the care of sick or wounded persons.

Hovel   An open structure used for sheltering cattle.

Hypaethral   A building having no roof or having an opening in the roof. It is open to the sky to admit light.

Hypocaust   An underground network of ducts below Roman ground floors to provide central heating. Warm air from a furnace circulated through these and travelled up wall ducts acting as flues.

Hypostyle   A hall or other large space, the roof of which is supported by rows of columns.

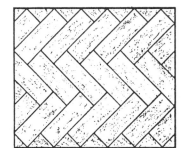

Herringbone

Imbon   To vault or arch over.

Impost   The point from which an arch springs. (5)

In-antis   *See* Anta.

Inband   Any stone or brick having its longest length built into the thickness of a wall e.g. a header brick, particularly a quoin.

Incertum   Random rubble masonry consisting of small stones bonded in mortar to irregular courses.

Infirmary   A hospital.

Inglenook   A seat recess built into a chimney breast or beside a fireplace.

Inlay   The sinking of contrasting strips of material flush into the surface of another material as an ornamentation. *See also* Marquetry.

Insula   Roman multistoried tenement blocks for servants/ slaves.

Intercolumniation   The distance between the shafts of adjacent columns measured in diameters or modules at their base. Vitruvius defined five main forms of intercolumniation: pycnostyle one and a half diameters apart; systyle two diameters apart; eustyle two and a quarter diameters apart; diastyle three diameters apart; araeostyle four diameters apart. (43)

Interdentils   The space or gap between adjacent dentils.

Intrados   The soffit and lower curve of an arch. *See also* Extrados. (5)

Ionic order   Greek and Roman order of architecture, characterized by its column capital which consists of four volutes.

Isodum   Masonry walls having courses of equal thickness.

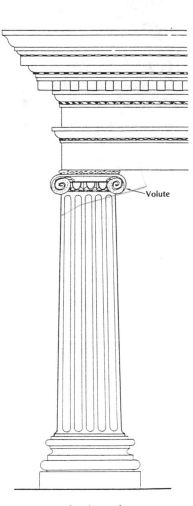

Ionic order

J

Jetty. A jetted timber frame

Jacobean Renaissance architecture during the rein of James I. When applied to furniture it means of a dark oak colour.

Jetty The part of a building that projects beyond the rest and overhangs the wall below.

Jib door A concealed door that is flush with the wall surface and decorated to provide a continuity of surface.

Jones Inigo A Renaissance architect who introduced Palladian style architecture to England.

Jutting Overhanging, a jetty.

K

Keel moulding A rounded moulding finishing in a pointed edge, like the keel of a boat.

Keep The donjon or main tower of a castle.

Keystone The highest central stone or voussoir of an arch or ribbed vault. (5)

King post The central vertical post of a roof truss between tie and ridge.

Kiosk An open pavillion supported by columns or pillars. Used mainly in gardens for band stands etc., also small freestanding shops.

Kitchen A room in which to cook. *See also* Ewery.

Kneeler The sloping topped, level bedded stones at the top of a gable end. *See also* Springer.

Label   A hood mould or drip located over an opening.

Labyrinth   A series of maze-like interconnected passages normally below ground.

Lancet   An early English style, sharply pointed arch resembling a lance. (5, 32)

Lantern   A small structure crowning a roof or dome for light, ventilation or as an ornament. (23)

Larder   Originally purely a storeroom for meat.

Laundry   A room where clothing etc. is washed, dried and ironed.

Lavatory   Originally a trough or sink for washing purposes, now taken to be the small room containing the wash basin and WC pan, or even just the WC pan on its own. *See also* Garderobe.

Lay light   A roof or ceiling light.

Leaded light   Windows consisting of rectangular or diamond-shaped pieces of glass held in lead cames (strips).

Leaf and dart   An ovolo moulding similar to egg and dart, consisting of alternate leaf and dart patterns.

Library   A room or rooms used for the storage of books.

Light   An opening or space created between the mullions and transoms of a window. *See also* Day.

Linenfold   A carved Tudor panel that resembles a folded, hanging piece of linen.

List or listel   A fillet.

Lobby   A small room that connects with one or more other rooms or apartments. *See also* Vestibule.

Lodge   The workshop and living quarters for the stonemasons set up when major medieval buildings were constructed. Also used for a small house at the gate of an estate or any small house in a park.

Loft   The roof space formed between the rafters and the ceiling of the upper floor rooms.

Loggia   An open-sided gallery or arcade, usually in a garden.

Long and short work Saxon quoins or corners consisting of alternate long stones on end laid on top of flat, horizontal slabs all bonded into the rubble wall.

Long gallery The major first floor room of E and H plan shape Tudor and early Renaissance manor houses. It ran the length of the house and was used for exercise, music and conversation.

Loop or loophole A small narrow light or slit in a wall through which arrows and other missiles could be fired. Often cruciform with circular enlargements at the ends and middle. Also the series of vertical doors of a multistoried warehouse through which goods are delivered.

Louvre Originally the cover over the hole in the roof of a medieval hall which permitted the smoke to escape but prevented the entry of rainwater. Now used for any horizontal timber or glass slat used in a ventilator, light or door.

Lozenge A diamond shape, often used in decoration.

Lucarne A dormer window. Also any small opening in a spire or roof.

Lunette A semicircular opening, a semicircular surface or any semicircular feature.

Machicolations An overhanging battlement in a medieval fortified building, with openings between the supporting corbels to enable boiling oil and other unpleasant substances to be dropped on the attackers below. (9)

Masonry Walls or buildings constructed from dressed or undressed stone. *See also* Ashlar and Rubble masonry.

Manor house The house of the manor. An unfortified, late medieval house.

Mansion A large house, a manor house.

Mantel The wood, metal, brick or stone frame surrounding a fireplace. The beam over the opening is termed the mantel-tree and the shelf the mantelpiece. *See also* Overmantel.

Margin An undecorated uniform border worked around the edges of a member or panel. A margin light is a narrow pane at the edge of a window. (27)

Marquetry The decoration of a timber surface by inlaying into it thin pieces of metal, ivory, mother of pearl or contrasting wood veneers.

Mausoleum A stately tomb.

Medieval architecture Gothic or pointed architecture. European architecture during the Middle Ages.

Megalithic Constructed or consisting of large stones.

Menhir A large, tall upright stone.

Merlon The solid part of a battlemented parapet. (9)

Mesaulae Passages or small courts.

Metope The spaces between the triglyphs of a Doric frieze. These may be either plain or enriched. (23)

Mews A terrace of coachhouses or stables with living accommodation for coachmen or grooms over. Mainly associated with, and built at, the rear of large town houses.

Mezzanine A partial storey midway between two main floor levels. Also termed an entresol.

Middle Ages The period of Western European history beginning with the fall of the Roman Empire in the fifth

Coursed random rubble

Channel jointed rustication

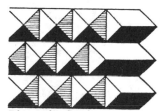

Diamond pointed

Masonry

century and ending with the Early Renaissance in the sixteenth century. Also termed the Dark Ages. *See also* Medieval architecture.

Minster The church which is, or was at one time, attached to a monastery.

Minute A unit of measurement in classical architecture, representing one-sixtieth of the diameter of a column shaft at its base. A subdivision of a module.

Moat A deep trench around fortified buildings, often filled with water.

Modillion A small projecting console bracket often used in pairs to support classical cornices.

Module A unit of measurement which controls proportion in classical architecture, representing either the diameter or half the diameter of a column shaft at its base. Minutes are subdivisions of modules. *See also* heading under General.

Monastery An establishment for monks.

Monolith A single column of stone, often a monument.

Monument A structure to commemorate a person or event. *See also* Triumphal arch.

Mosaic Ornamental floor and wall surfaces, formed by inlaying small pieces of glass, stone or marble in cement. Also termed tessellated.

Motte A manmade mound of earth on which the timber tower stood in a Norman motte and bailey castle.

Moulding The ornamental contours given to the angles and features of a building. Variously named according to their profile.

Mullion The vertical member forming the division between the lights of a window, or other opening. *See also* Transom.

Muntin The intermediate vertical framing member of a door or panelling which is jointed into the horizontal rails.

Mural A wall painting or hanging. *See also* Fresco.

Museum A building to house or exhibit a collection of objects of historical or scientific interest.

Mutules The projecting square blocks attached under Doric cornices. *See also* Orders. (23)

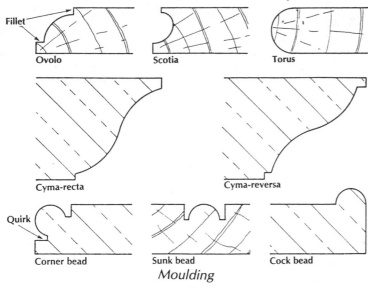

Moulding

Nail head An early English ornamentation consisting of a band resembling a row of projecting nail heads.

Nash A Regency architect whose most famous buildings include, Buckingham Palace, works in Regent's Street and Regent's Park, Marble Arch and the Brighton Royal Pavillion. Nash favoured the use of stucco to cover the whole façade of a building. A rhyme of the time was: 'Oh Nash was he a great master he found us all brick and left us all plaster'.

N

Nave The part of a church in which the congregation assembles.

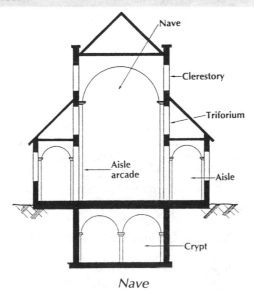

Nave

Niche A recess in a wall or pier for a vase or statue etc. Often arched over and semicircular in plan.

Norman architecture The style of architecture introduced by the Normans who conquered England in 1066. Characterized by stone-built manor houses with undercroft and solar; stone keep and bailey castles; churches with massive, thick stone walls, large rounded columns, semicircular headed arches, doors and windows; vaulting and squat square towers.

Obelisk A tall, tapering stone shaft normally square in plan with a pyramidal top.

Observatory A building or place where astrological movements can be studied.

Octastyle A portico or colonnade with eight frontal columns.

Oculus A circular opening in a wall, sometimes blind.

Offset The projection from the face of a wall where it increases in thickness. Where these are horizontal as when the wall above is reduced in thickness they should be weathered and have a projecting drip.

Ogee A double curve S-shaped line or moulding known as a cyma recta. (5)

Orangery A glazed garden building for growing oranges. *See also* Greenhouse and Conservatory.

Oratory A small private chapel in either a private house or church.

Orders In classical architecture an assembly consisting of a base, column capital, entablature and pediment decorated and proportioned according to one of the five orders: Doric, Ionic, Corinthian, Tuscan and Composite. There are Greek and Roman versions of the Doric, Ionic and Corinthian, while Tuscan and Composite are Roman additions. Ionic capitals have large volutes; Corinthian capitals consist of acanthus leaves; Greek Doric columns have no base; the Tuscan column has an unfluted shaft; and the Composite order is a mixture of the Ionic and Corinthian.

Oriel A balcony, platform or window that projects from an upper storey and is carried on brackets or corbels.

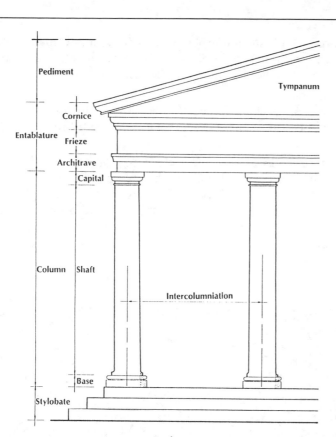

Orders

Orientation   The siting of a building in relation to the points of a compass or the rising sun.

Ornament   The enrichment or decoration of mouldings and other details for aesthetic purposes. They do not form an essential part of the construction.

Orthostyle   Having columns in a straight line.

Oubliette   A small, secret prison cell reached only by a trap door in the ceiling above.

Overmantel   The framework above a mantelshelf often containing a painting or mirror.

Oversail   An overhanging portion of a building. *See also* Jetty and Jutting.

Ovolo   A convex moulding much used in classical architecture consisting of a quarter round with a fillet on either edge. Roman versions are formed from arcs of a circle, while Grecian ones are flatter, being based on elliptical curves. *See also* Egg and dart. (41)

P

Pagoda An eastern temple in the form of a tall, many storied tower with ornamental roofs at each level. Used as an chinoiserie eyecatcher in the eighteenth century.

Palace An official dwelling of an archbishop or bishop. A spacious stately mansion. Also a royal household.

Palladian architecture A style based on the buildings and publications of the Italian architect Palladio. Introduced to England by Inigo Jones. Characteristics of the style are: symmetric design; a roof line that does not project above the walls; pediments over windows, doors and all principal rooms on the first floor e.g. piano nobile.

Palladian architecture

Pan The space between vertical posts in timber-frame buildings. *See also* Pane.

Pane The pan of a timber-frame building that contained a window. Hence window pane.

Panel A distinct area that is raised above or sunken below its general surroundings or framework as in panelling, doors and ceilings. *See also* Field and Margin.

Parapet A low wall to provide projection from a sudden drop e.g. at the eaves of a roof or tower etc. It may be plain, battlemented or ornamented. *See also* Attic.

Parclose A screen enclosing a chapel or tomb to separate it from the main body of the church.

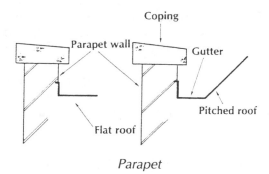

Parapet

Pargeting Decorative patterned plasterwork used to finish external walls. Also the plaster lining of chimney flues.

Parlour A room for conversation.

Parquet Highly polished hardwood flooring formed using brick size blocks, often laid in a herringbone pattern.

Parsonage A building used as a priest's dwelling.

Passage A corridor providing access between the different rooms of a house. Also a narrow pedestrian street in a built-up area. *See also* Screens passage and Alley.

Patera A small, flat circular ornamentation resembling a rose.

Patio An atrium or inner open courtyard.

Pavement A path or road laid with stone or other material e.g. cobbles, setts, bricks, flags, concrete or Tarmacadam.

Pargeting

Pavilion A lightly constructed building, often ornamental. Used as a garden summerhouse; or as a sports changing room (cricket pavilion); or attached as a wing to a main building. A pavilion roof is one that has equal hips on all sides.

Pedestal The structure or stand that is used to support some columns and also statues and vases. Consisting of a plinth, dado and cornice. (33)

Pediment A classical, low-pitched, triangular gable end supported on the entablature above a portico, often containing a coat of arms or other sculptured decoration. Also the similar feature above doors, windows and other openings, where they may be triangular or segmental. Pediments over openings are sometimes open to either the apex or base. *See also* Aedicule and Tympanum. (43)

Pendant   An ornamentation suspended from the ribs of a vault or a stucco ceiling. Also any member may be termed pendant where it projects or is suspended from a soffit.

Pendentive   A triangular curved section which forms the transition from the circular base of a dome or its tambour on to the square or polyagonal supporting structure. *See also* Squinch. (23)

Penthouse   Any subsidiary building with a monopitch or lean-to roof. Also a separately roofed structure on the roof of another building.

Pergola   A covered garden walkway consisting of a double row of posts spanned by beams overhead and covered with climbing plants.

Peristyle   A building or open court surrounded by a continuous range of columns.

Perpendicular style   The final variation of Gothic pointed architecture. Characterized by straight, slender vertical elements; tall windows subdivided by mullions and transoms; restrained tracery; ribbed and fan vaulting; extensive use of internal panelling. (33)

Perron   The flight of stairs and landing that lead up to the main entrance of a building normally at first floor level. *See also* Piano nobile.

Pew   Fixed wooden church seating.

Piano nobile   The main floor of a building containing the living or reception rooms. In classical architecture this is the first floor. *See also* Perron.

Piazza   An open square or rectangular space surrounded by buildings.

Picture gallery   A building or part of a building used to display drawings and paintings.

Piedroit   A pilaster having no base or capital.

Pier   A solid mass of masonary between doors, windows and other openings in a building. *See also* Pillar.

Pilaster   A shallow, rectangular pier that projects from a wall surface and conforms to orders.

Thickened pier for stability

Main wall

Pier

Pillar   A freestanding, vertical pier not conforming to orders. A column without a base and capital.

Pilotis   A building standing on pillars so that the ground floor is open.

Pinnacle   The cone or pyramid-shaped termination of a buttress, spire or parapet, often decorated with crockets.

Plafond   Any ceiling or soffit.

Plaisance   A summerhouse near a mansion.

Planted   A moulding that is worked on a separate piece and then fixed in position, as opposed to stuck mouldings which are worked on the solid component.

Plate tracery   The earliest form of tracery from the early English period where the spandrel between the dripstone and lancet windows was pierced with circular trefoil and quatrefoil holes.

Plinth   The projecting base of a pedestal or wall. A plinth block is the plain block at the base of a door architrave. Also the projecting or recessed base of a cupboard or other similar structure. *See also* Skirting. (48)

Podium   A continuous pedestal.

Pointed arch   A Gothic arch or any arch with a pointed head.

Polystyle   A building or structure having many columns.

Pommel   A globular ornament used to terminate a pinnacle.

Porch   An exterior projection over a doorway forming a covered approach. Its sides can be either open or closed. Those having columns may be termed porticos. *See also* Porte-cochere.

Portcullis   A reinforced defensive grating at the entrance to castles and other fortifications. It is raised or lowered vertically by sliding in a groove or cullis.

Porte-cochere   A porch capable of admitting vehicles.

Portico   A porch to the entrance of a building which is supported by a range of columns. A feature of Renaissance architecture often pedimented. *See also* Orders. (56)

Post   Any vertical piece of timber. *See also* Pan.

Postern   A private entrance.

Potboard   The lowest board or shelf of a cupboard, historically named after a low board on which pots were stored.

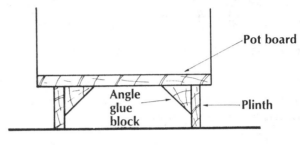

Potboard

Prison   A building where persons are kept in captivity. *See also* Oubliette.

Profile   The section of a moulding or the outline of a building.

Prostyle   A projecting portico having freestanding columns in a row, not in antis. *See also* Anta. (56)

Puncheon   A short vertical post or framing member e.g. a stud. *See also* Stanchion.

Pycnostyle   One of the five styles of intercolumniation having an arrangement of columns spaced one and a half shaft diameters apart. *See also* Araeostyle, Diastyle, Eustyle, and Systyle.

Pyramid   An Egyptian monument having a square or polyagonal base and sloping sides meeting at the apex. Also used for any structure or detail taking a similar form.

Quadrangle   A square or rectangular courtyard enclosed on all sides by buildings.

Quarrel or quarry   In leaded lights a small diamond shaped piece of glass or a square one positioned diagonally. Also used for any small square shaped component e.g. floor tiles.

Quarter round   An ovolo moulding.

Quatrefoil   A tracery pattern consisting of four foils. *See also* Cusp.

Quay   A river or sea bank formed to enable the loading and unloading of boats.

Queen Anne   A style of architecture for smaller domestic houses, pioneered by Sir Christopher Wren who successfully combined the palladian style with English features such as steeply pitched roofs, large windows and tall chimney stacks.

Queen posts   Two vertical posts placed symmetrically on a roof tie beam to form a truss.

Quirk   A recessed, narrow groove alongside a mould. (41)

Quoins   The external corners of a building. Also refers to the bricks or stones used to construct an external corner.

Rampant arch   An arch having its springings on different sides at different levels.

Rampart   A stone or earth wall used for defence purposes.

Reed or reeded   A moulding consisting of parallel, convex mouldings or beads touching one another. *See also* Flutes.

Refectory   A dining hall.

Regency   The last stage of the Georgian period when the Prince of Wales, was Regent during the last years of George II's reign.

Regrating   Also known as skinning. The removal of the surface of old stonework.

Relief   The projection of any ornament or sculpture from its background.

Regency

Relieving arch   An arch built over a lintel or beam to take off some of the load, thus preventing lintel collapse.

Renaissance   The rebirth or rediscovery of classical style which is based on symmetry and the concept of orders, each part being mathematical in proportion and related to each other. *See also* Golden section.

Reredos   A carved wooden or stone screen positioned behind an altar.

Return   The continuation of a moulding or projection in another direction.

Reveal   The part of a door or window jamb between the frame and the wall surface.

Revetment   A retaining wall built to hold an earth bank.

Rib   A projecting band on a vault or ceiling surface, usually for structural purposes, but can be totally ornamental.

Rococo   Part of the Baroque style of Georgian architecture. Where ornate decoration of gold leaved stucco or carved plants, shells and garlands was applied to architecture and furniture.

Roll   A rounded piece of timber used for dressing lead over. A roll moulding is a rounded Gothic mould also called bowtell or astragal.

Roman architecture   The development of the classical style of architecture belongs to the Greeks who used a column-and-beam, trabeated form of construction. On colonizing Greece, the Romans incorporated the trabeated orders into their arch and dome arcuated buildings, although often as merely applied decoration. The structural use of arch and dome made great engineering works possible, many of their buildings being four and five storeys high. In England from AD 43 they built roads and walled towns on a regular grid pattern, each containing domus, insula, shops, baths, temples, arenas and town halls. In addition they also had central heating and efficient water supplies and drainage systems. *See also* Hypocaust.

Romanesque   The period between the break-up of the Roman Empire and the coming of the Gothic era. Mainly applied to the Anglo-Norman period in England between 1066 and 1200.

Rose window A circular tracery window used in Romanesque and Gothic churches.

Rosette A rose-like ornamentation. *See also* Patena.

Rostrum A raised platform or dias.

Rotunda A circular building often peristyle. Also a circular domed room.

Roundel A circular mould or panel.

Rubble masonry Walls constructed of rough, undressed stones. Random rubble is uncoursed. Coursed rubble is roughly dressed stones built to occasional courses.

Rustication The working of deep grooves or channels in the joints of masonry walls, employed to emphasize the lower part of external walls. Sometimes also simulated in stucco. (39)

Sarcophagus A stone coffin.

Saxon architecture The style of architecture introduced by the Saxons who invaded England in AD 410. Being a seafaring nation they applied their boat-building skills to the construction of cruck frame cotes and halls. Later they built stone churches with long and short quions and ornamental pilaster strips, lattice work and blind arcading.

Scallop An ornament in the form of scallop shells.

Scotia A hollow or concave moulding which is formed by two quadrants of different radii. Often used on the base of a column between two torus mouldings. *See also* Cavetto. (41)

Screen A partition to separate one portion of a room from another.

Screens passage Passage at the service end of a medieval hall formed by a screen concealing the main entrance, also the pantry, buttery and kitchen. Often the screened area was floored over to create a minstrels gallery overlooking the hall.

Scroll A spiral ornament taking the form of a partly rolled piece of paper e.g. a volute in classical architecture.

Scullery The room where dishes are washed. *See also* Ewery.

S

Sedilia Recessed seats in a wall of the chancel.

Serpentine A wall having a wavy plan shape; also known as crinkle-crankle.

Sett A cubicle piece of stone used for a pavement.

Setts

Shaft The main body of a column between the base and the capital. (43)

Shamble or shambles An abbattoir or place where animals are killed.

Shelf A board fixed to a wall used to carry or display objects. *See also* Cupboard and Potboard.

Shell door canopy A canopy in the shape of a shell hung over an external door for protection and decoration.

Sheraton A Georgian cabinet maker.

Shroud A church crypt.

Shutters Internal or external doors or frames used to cover windows.

Skinning *See* Regrating.

Skirting The board fixed at the junction between the internal wall and floor, externally termed a plinth. *See also* heading under Building construction.

Skylight A window set into a roof or ceiling to provide natural lighting from above. *See also* Lantern light.

Shell door canopy

Soffit   The underside of a building element, such as a ceiling or the lower surface of a beam, arch or vault. (5)

Solar   The withdrawing or private room of a Norman manor house where the Lord and his family could withdraw for privacy and to sleep.

Solarium   A sun terrace, room, or loggia.

Solomonic column   A barley sugar twisted column.

Spandrel   The triangular space between the curve of an arch and the square enclosing it or any approximately triangular area.

Spire   A very tall, pointed structure forming the roof of turrets and towers, a typical feature of Gothic churches.

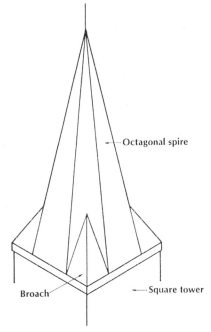

Spire (broach octagonal spire from a square tower)

Spital   A hospital.

Splay   A sloping surface, often applied to window and door reveals that have slanting sides. *See also* Chamfer.

Springer   The overhanging stone at the corner of a gable end below the lowest kneeler.

Springing   The impost of an arch, the point at which the curved surface meets its vertical support. (5)

Square   A paved or planted open area having four equal length sides and right-angled corners. *See also* heading under General.

Squinch   A small arch built diagonally across an internal corner in order to support a round or polyagonal structure (domes and spires) on a square plan.

Squint   An opening cut through a church wall or pier in an oblique direction to allow people in the aisles a view of the main altar.

Stable   A building used to accommodate horses.

Stadium   A sports ground for athletic activities.

Stall   One wooden or stone carved seat in a row of similar seats. *See also* Stallboard under Building construction.

Stanchion   A main vertical supporting member or prop. *See also* Puncheon.

Steeple   The tower and spire of a church.

Stoep   A verandah.

Story or storey   The space between two floor levels of a building e.g. ground storey or floor, first storey or floor etc.

Strapwork   A form of decoration using intertwined, raised bands of timber or plaster forming a fretwork. Applied to furniture, doors, screens and plaster ceilings in the Elizabethan and Jacobean periods.

Striated   A channelled surface.

String course   A recessed or projecting continuous band or moulding on an external wall.

Stucco   Plasterwork either lime based and rendered smooth to coat external walls, or fine gypsum and marble dust based and used for interior moulding and ornamentation.

Stylobate   The raised, stepped structure in classical architecture that supports a colonnade. Strictly only the top step. (43)

Subway   An underground passage used mainly by pedestrians. Also for sewers and other services.

Summer   A beam, lintel or breastsummer.

Surbased An arch, dome or vault rising in height less than half its span.

Swag An ornamentation consisting of a festoon of fruit, flowers or cloth held on two supports.

Swan neck Any member having a double curve form resembling a swan's neck e.g. a swan neck handrail, a swan neck rainwater pipe.

Symmetry The uniformity or balance of one part of a building with the other – either side of the centre point must be an identical mirror image. Forms the basis of classical architecture.

Systyle One of the five styles of intercolumniation having an arrangement of columns spaced two diameters apart. *See also* Araeostyle, Diastyle, Eustyle and Pynostyle.

Tabernacle A freestanding canopy or canopied recess.

Tambour The circular wall that carries a dome or cupola. (23)

Template A setting out pattern.

Temple A place of worship.

Terrace A raised platform or space adjacent to a building. *See also* heading under General.

Tessellated A mosaic floor or wall covering, made by embedding glass, stone or marble cubes in cement.

Terrace

Plan of a prostyle tetrastyle portico

Tetrastyle   A portico or colonnade with four frontal columns.

Thatch   A roof covering of reed or straw, tied together with flexible willow sticks.

Theatre   A building or place for dramatic or musical public performances.

Thermae   A Roman public bath.

Thermal window   A feature of Palladian architecture. Also termed the diocletian window.

Timber framing   Half timbered construction where walls are constructed of vertical and horizontal timber members. The spaces between are filled with a non-structural infill.

Tomb   A grave including any monument.

Tooth ornament   *See* Dog tooth.

Torus   A large, convex, semicircular moulding, often used at the base of a column. *See also* Scotia and Astragal. (41)

Tourelle   A turret corbelled out from the face of a wall.

Tower   A tall building or structure, used for defence; a watchtower; a landmark or in churches for the hanging of bells. Often topped by a spire. *See also* Steeple.

Trabeation   A building that uses the post and beam principle as its form of construction as opposed to the arch or arcuated method.

Tracery The ornamental working of Gothic mullions and transoms used in windows, screens, panels and vaults. *See also* Plate tracery and Foil.

Transept The arms of a cross-shaped church set at right angles to the nave and choir.

Transition A term used to describe buildings in which the change from one architectural style to another is clearly visible, especially the transition from Romanesque to Gothic.

Transom The horizontal member forming the division between the lights of a window, or other opening. *See also* Mullion.

Trefoil A tracery pattern consisting of three foils. *See also* Cusp. (5)

Triforium An arcaded gallery or area under an aisle roof, situated between the nave arcade and the clerestory windows. (41)

Triglyph The vertical fluted panel in a Doric frieze situated between the metopes. (23)

Triumphal arch A monumental arch built to celebrate an event or person.

Tudor style The transitional architectural period between perpendicular Gothic and Renaissance. Characterized by the building of many E and H plan shaped palaces and mansions, using brick which had recently been introduced from Flanders; tall decorative patterned chimneys; diapering and black and white work. (5)

Tudor style

Tunnel An underground channel covered by a vaulted roof.

Turret A very small slender tower or a large pinnacle.

Tuscan order A column with an unfluted shaft.

Tympanum The triangular face of a pediment enclosed by its mouldings. Also the area between the lintel of a doorway and a relieving arch over it. (43)

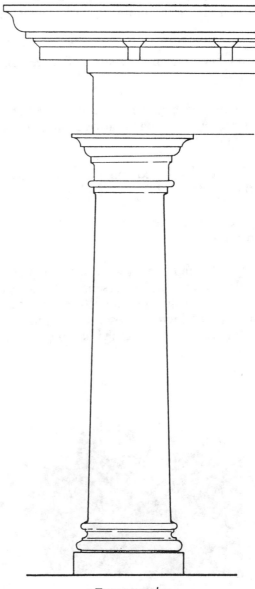

Tuscan order

Undercroft   A vaulted room under a room above which may be above or below ground level. *See also* Crypt.

Vault   An arched ceiling or roof over a room constructed of brick or stone. A barrel vault is a simple, semicircular tunnel. A cross groined vault is where two barrel vaults intersect at right angles. A rib vault is one having ribs that divide the bay and carry the infilling stones between them. A fan vault has a fan-like springing of ribs. *See also* Pendant.

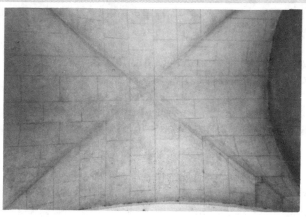

Vault

Venetian door   An arched central door opening, with narrow flat-topped windows on either side.

Venetian window   A main central window, often arched, flanked by narrow, flat-topped windows on either side. Also termed a Palladian window.

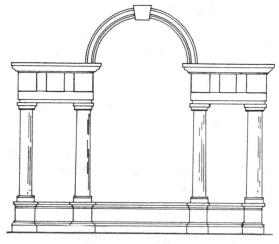

Venetian window

Verandah An open gallery or balcony on the outside of a building. Having its own roof or canopy normally supported on light pillars. Also termed a stoep.

Vermiculated The decoration of stonework surfaces with irregular shallow channels. Has the appearance of being eaten by worms or larvae.

Vesica An upright, pointed, elliptical shape used in decorative panels to enclose a figure of Christ enthroned.

Vestibule An entrance lobby or hall from which doors open into the various rooms of a building. Also an open court in front of a building.

Vestry A room adjoining a church in which the vestments (clergy and choristers' garments) are kept.

Viaduct A series of arches forming a bridge to carry a road or railway. *See also* Aqueduct.

Victorian architecture The Victorian era from 1837 to 1901 saw a massive building expansion. More buildings were constructed at this time than in all the previous eras put together. The Industrial Revolution was accelerating rapidly. An extensive railway system, a canal and road network, factories, offices, shops, houses, educational, public and civic buildings were all developed. The period saw the development of the Gothic revival which had started during the Regency period, resulting in the battle of styles between Gothic and classical architecture. Domestic architecture was mainly Gothic revival: spires, turrets, battlements, diapering, stained glass leaded lights, mosaics and cast iron gates and railings. Other events and features of the time were the 1851 exhibition at Crystal Palace, back to back housing, the introduction of electric lighting and telephones, the Public Health Act, tap water and the increasing provision of lavatories. *See also* Gothick.

Villa A Roman landowner's residence, a country house, a detached or semidetached house on the outskirts of town.

Vitruvius A Roman architect whose writings had enormous influence on Renaissance building. His writings on architecture which were rediscovered and reprinted in the fifteenth century became essential to all progressive architects.

Volute A spiral scroll or helix forming part of a column capital. *See also* Orders.

Victorian

Voussoir   A wedged shape brick or stone forming part of an arch. (5)

Wagon ceiling or wagon vault   A barrel vault.

Wainscot   Timber wall panelling or lining, originally oak.

Ward   The bailey of a castle.

Wattle and daub   The infilling between the framework of half timbered buildings and cruck framed cotes and halls. Consisting of a woven basketwork of timber lathing called wattle, plastered or daubed over with a mixture of clay, animal fat and cow dung.

Wealden   A medieval half-timbered house with open hall, flanked by two-storied, jettied end chambers. The eaves being continuous had a deeper projection over the hall block and were supported by curved brackets.

Weathercock   A vane fixed to a building at a high level to indicate wind direction.

Weathering   A slope given to an external horizontal surface in order to throw off rainwater.

Wicket   A small door to provide pedestian access through a larger door or gate. Also the gate in a dwarf partition.

Window tax   A tax levied between 1697 and 1851 on houses with more than six glass windows resulted in many windows being bricked up to avoid payment. *See also* Blind.

Wings   The side projecting portions or subsidiary structures of a main façade.

Woods   A father and son team of Georgian architects who designed much of Bath.

Wren, Christopher   A Renaissance architect who designed St Paul's Cathedral in London. Also, as architect to Charles II, was responsible for rebuilding the fifty-one London churches destroyed in the Great Fire of 1666.

St Paul's cathedral designed by Christopher Wren

St Paul's internal decoration

Y

Yard   A paved area, often at the back of a house and enclosed by a wall.

Yorkshire bond   Monk bond. See also heading under Building construction.

Yorkshire light   A pair of lights, one fixed, the other sliding horizontally.

Z

Zig-zag   A chevron decoration used in Norman buildings consisting of mouldings running in zig-zag lines.

Zig-zag decoration around a Norman doorway

2
BUILDING
CONSTRUCTION

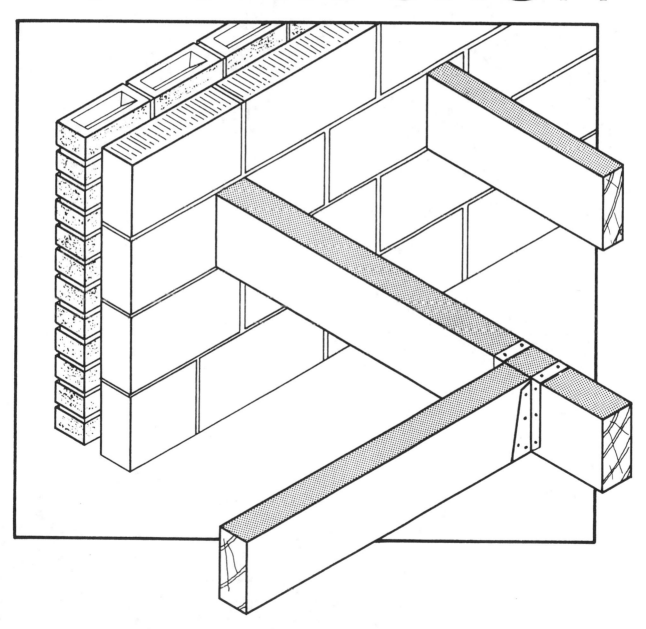

Abut   To join or meet. *See also* Butt.

Abutting tenons   Two tenons that enter from opposite edges of a member and meet in the middle of the mortise. *See also* Mortise and Tenon.

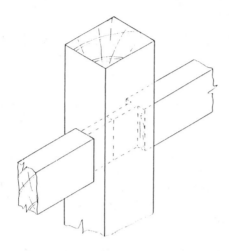

Abutting tenons

Access   The means (doors or stairs etc.) by which entry is made to a building or room. Also the means (panel, door or eye) by which entry is made to a building service for maintenance or repair.

Acoustic construction   Any building method that reduces sound cither entering, leaving or transferring through a structure e.g. by means of dense construction; use of absorbent materials or its discontinuous construction. *See also* Acoustics under Materials and scientific principles.

Adjustable steel prop   A proprietary prop with provision for adjustment so that its length may be varied for levelling and striking. Used either as vertical supports for slab formwork or an inclined member to hold vertical formwork plumb. Often abbreviated to ASP.

Air supported structure   A surface structure supported by compressed air or inflated tubes used for warehouses, sports complexes and exhibition areas.

Airbrick   A perforated brick or grid built into a wall to ventilate a room, cupboard or the space under a timber ground floor.

American bond   An English garden wall bond.

Apron lining   The board used to finish the edge of a trimmed opening in a floor. *See also* Stair. (149)

Arch   A curved structure bridging an opening and capable of supporting a load. Built of separate components each supporting one another by mutual pressure. Variously named according to shape.

Arch brick   Also termed a voussoir. A wedged-shaped brick for arch construction formed by axing, sawing or rubbing.

Arch centre   The structure, often timber, used to provide temporary support for an arch under construction. *See also* Turning piece.

Architrave   The decorative trim that is planted internally around door and window openings to mask the joint between wall and timber and conceal any subsequent shrinkage and expansion. *See also* Plinth block, Skirting and heading under Architectural style.

Arris   The sharp edge or corner of a building component. (105)

Ashlar   Dressed stone block mainly used as a facing fixed to a concrete or brickwork backing. *See also* heading under Architectural style.

Ashlaring   The vertical studwork between the rafters and ceiling joist forming the walls of rooms in the roof space. *See also* heading under Architectural style.

ASP   An adjustable steel prop.

Attic   A loft. *See also* heading under Architectural style.

Axed arch   An arch built using bricks that have been cut with an axe or bolster and not rubbed.

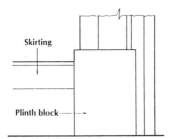

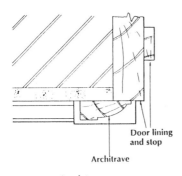

Architrave

Back filling   The refilling of drain and foundation trenches with earth, hardcore, or concrete etc. after the construction work has been carried out.

Back prop   The repropping of concrete beam and slab soffits after striking of formwork until the concrete gains sufficient strength to become self-supporting.

Back putty or bed putty   The putty or other glazing compound that remains between the glass and rebate. *See also* Face putty.

Back shore   The short outer shore in a multiple raking shore that provides support for the rider shore. (136)

Backing   The brickwork or stonework that is covered by facework. Any coat of paint or plaster other than a finishing coat. *See also* Backing bevel.

Backing bevel   The shaping applied to the top edge of a hip rafter to suit the two sloping roof surfaces.

Backlining   The thin board that is used to close the back of a box frame window in order to prevent the weights fouling on the brickwork. *See also* Wagtail. (73)

Bagging up   The making good of a concrete surface or brick joints by rubbing over the surface with mortar contained in a cement bag or piece of hessian sacking.

Balanced step   A dancing step.

Balcony   *See* heading under Architectural style.

Balloon framing   A form of timber-frame construction where the studs are continuous from the damp proof course level up to the eaves with intermediate floor joists being supported on ribbons. *See also* Platform frame.

Baluster   The vertical in-filling members of a balustrade. *See also* Banister and Stair. (149)

Balustrade   The handrail and infilling guarding the open edge of a stair, landing or floor. Can be termed open or closed depending on the infilling. No opening to permit passage of 100 mm sphere (prevents children's heads getting stuck in the gap) where the stairway is likely to be used by children. (141, 149)

Banding   Inlays or strips to cover the edges of veneers, often of a contrasting colour.

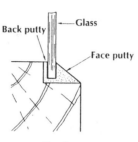

Back putty

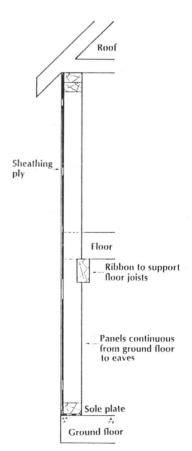

Balloon frame

Banister   A baluster, but also sometimes used to describe the entire balustrade.

Barefare tenon   A tenon having a shoulder on one side only, the other being flush with the timber face. Used for stair strings and framed ledged, braced and matchboarded doors.

Bareface tongue   A tongue formed by a rebate on one side only, the other side being flush with the timber face. Often used as the corner joint for door linings.

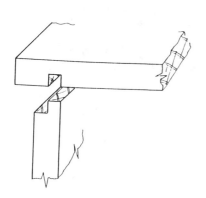

Bareface tongue

Bargeboard   The continuation of the fascia board around the sloping verge of a pitched roof. *See also* Finial.

Bat   A brick cut to reduce its length in order to form a bond. A slab of insulating material often built in cavity walls. *See also* Half bat. (72)

Batten door   A matchboarded door.

Battening   Fairly small sectioned lengths of timber fixed to a surface, in order to provide a flat ground onto which other members may be fixed e.g. tile battens. *See also* Grounds and Brandering.

Batter   A steep slope e.g. the sides of a foundation trench may be battered in order to prevent the earth slipping.

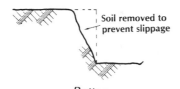

Batter

Batterboard   A profile board.

Bay   A division of a framed building e.g. the area enclosed between four columns. Also the area of concrete, or waterproofing cast, layed or screeding applied at one time. Also termed a panel.

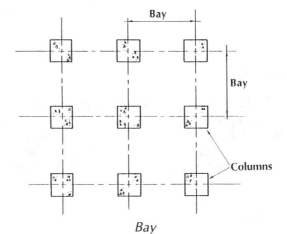

Bay

Bay window   A window that projects past the main line of a building and is carried on either a dwarf wall or brackets; variously named according to plan shape e.g. square, cant (splayed sides) and segmental. A bay that projects from an upper storey only is termed an oriel window. *See also* Bonnet.

Bead   A small moulding or trim that masks a joint. *See also* Glazing bead.

Beadbutt   A door panel that has a bead moulding stuck on its vertical edges, while its horizontal edges are left square and butt up to the rails.

Beam   A horizontal structural member spanning between supports.

Beam box   The formwork for the sides and soffit of a beam.

Beam filling   The brick or stone infilling between floor, or ceiling joists and rafters at their supports. Also termed windfilling.

Bearer   A horizontal timber that carries other timber members. (78)

Bearing bar or plate   A metal bar laid on a brick course to provide a strong level support for timber joists.

Bed   The underside of a building component e.g. brick, stone, tile, slate, wall plate or bearing bar.

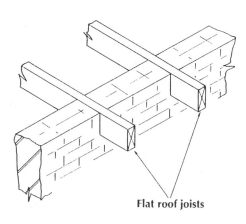

Beam filling

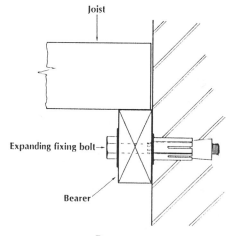

Bearer

Bedding or bed joint The layer of mortar, concrete or putty etc. on which the bed surface of a building component rests.

Belfast truss A timber or steel roof truss having a segmental top edge, also termed bowstring truss. Consists of curved upper chords, horizontal ties or strings and diagonal braces.

Bellcast eaves Sprocketed eaves.

Belt A string course.

Bench mark An ordnance bench mark used for establishing levels.

Bevel A right-angled corner at the meeting of two surfaces which has been removed asymmetrically. That is, more from one face than the other, unlike a chamfer which is symmetrical at forty-five degrees. Also used for any meeting of two surfaces which are not at a right angle. *See also* Cant, Splay and Tilted. (79)

Binder A timber or metal beam used in double floors to provide intermediate support for bridging joists. A horizontal timber used in pitched roofs and suspended ceilings, fixed across the top of the ceiling joists to provide intermediate support and thus stiffen them. Also a material that is used to bind together a mixture such as cement in concrete or mortar.

Birdsmouth The cutout near the foot of a rafter enabling it to fit over the wall plate. This is a combination of a plumb cut and seat cut.

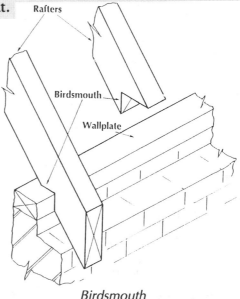

Birdsmouth

Blank A wall without door or window openings. *See also* Door blank.

Blank door or window A door or window opening that has been bricked up. Also termed blind. *See also* Door blank.

Blank wall A wall without any door or window openings.

Blind An item not seen or one which cannot be seen through.

Blind mortise A stopped mortise that does not completely pass through a piece of timber, used to accommodate a stub tenon.

Blind nailing Secret nailing.

Block bonding The bonding of several brick courses of one wall into another, thus each course is not bonded separately. *See also* Toothing.

Blockwork Walls constructed from blocks.

Blowing The expulsion of areas of paint, plaster or rendering from its backing. *See also* Popping.

BM A bench mark.

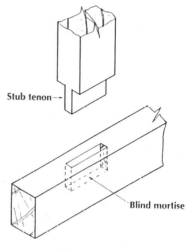

Blind mortise

Block bonding

Boarding Timber boards laid over a surface. *See also* Close boarding and Cladding.

Bond The overlapping of vertical joints in adjacent courses of bricks, blocks, tiles and slates to evenly distribute loading or for weathering. Bonds variously named according to the amount of overlap and the arrangement of components. Also the adhesion between two materials as in an adhesive or bonding agent.

Bond course A course of headers.

Bonding The laying of brick, blocks, tiles and slates with overlapping vertical joints. It increases the strength of brick and block walls by distributing the load throughout the wall, and the weather resistance of tiles and slates. Also a type of plaster used to bond to concrete.

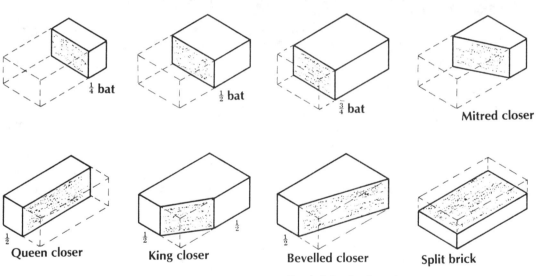

Cut bricks for bonding

Boning rods Tee-shaped devices used to sight-in the line of drainpipes and foundation trenches etc. Used in sets of three, the middle one being termed a traveller.

Bonnet Hip tiles or capping that are bedded in mortar, one overlapping the other. Also the roof to a bay window.

Boot The projecting nib on a lintel, beam or floor slab to carry the face brickwork above.

Borrowed light A light or window in an internal wall or partition. Said to borrow daylight from an adjacent room with external windows. (141)

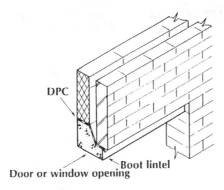

Boot

Bottom rail   The bottom horizontal member of a door, casement or lower sash. (77)

Bowstring truss   A Belfast truss.

Box frame   The traditional pattern of a sliding sash window, where the sashes are counterbalanced by weights contained in a boxed framework.

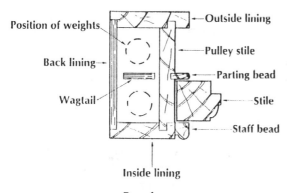

Box frame

Boxing in, out or up   The encasing of a building element or component with timber.

Brace   A structural member used to triangulate rectangular frameworks in order to stiffen them. Struts are mainly in compression and ties mainly in tension. (87)

Bracket   A projecting support member, either solid or with an inclined brace.

Bracketed stairs   A cut stair string with ornamental brackets fixed under the overhanging return nosings. *See also* Carriage piece.

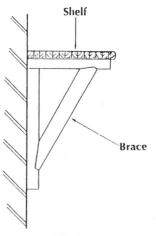

Bracket

Bracketing   Timber brackets used to support a plaster cornice from a wall or ceiling. Also the cradling around steel beams.

Brandering   *See* heading under Architectural style.

Breast   *See* heading under Architectural style.

Breastsummer   *See* heading under Architectural style.

Brick construction   Having load bearing brick walls.

Brick nogging   A non-load bearing brickwork infill between vertical structural members. *See also* Nog brick.

Brick on edge   A sill or coping of headers laid on edge. *See also* Soldier course.

Brick skin or brick veneer   Load bearing timber-frame construction where brickwork forms the external weathering and decorative layer.

Brickwork   Walling built in courses, laid to a bond, and bedded in mortar.

Bridging   The spanning of a gap in a building element.

Bridging joists   A floor joist that spans from support to support, also termed a common joist. *See also* Binder and Trimming joist.

Bridle   A trimmer joist.

Building component   *See* Component.

Building element   *See* Element.

Building-in   *See* Built-in.

Building line   A line established by the local authority in front of which no building can be erected. This is normally taken as the outside face of a building's wall. Its purpose is to ensure buildings are set back sufficiently from adjacent roadways. *See also* Improvement line.

Built-in   A component fixed into a wall or other element by bedding in mortar and surrounded with the walling components e.g. an air brick, cavity tie or floor joist. Joinery that is specified as built-in means items that are inserted during the main building process rather than fixed-in, which are inserted later.

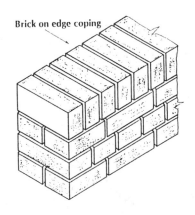

Brick on edge coping

Brick on edge

Built-up centre   A centre for an arch that is made up from a number of pieces, rather than a turning piece which is cut from the solid.

Bulkhead   The sloping ceiling above a stair to permit sufficient headroom. Often associated with the trimmed stairwell opening. May incorporate a cupboard.

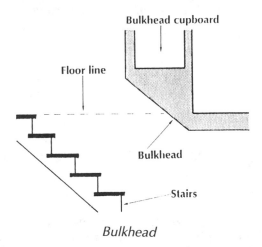

Bulkhead

Bullnose step   The quarter-rounded end step that projects past the face of the newel, often located at the bottom of a flight of stairs. *See also* Commode step and Curtail step. (149)

Butt   To join without overlapping. *See also* Abut and Abutment.

Butt joint   A joint formed between two materials of the same thickness, normally timber or metal, which meet at their ends or edges without overlapping or penetrating each other.

Buttering or buttering up   The spreading of mortar on the vertical joint of bricks before laying. Also the application of adhesive to the back of a wall tile before pressing into place.

Button   A device used to secure solid timber table and worktops to their supporting framework or carcass while still allowing an amount of moisture movement. Splitting of the top could result if the top were to be held tight.

Buttress   *See* heading under Architectural style.

C

Caisson A box-like structure sunk below ground or water level. This is cleared of soil or pumped free of water to provide a protected area for construction work to take place.

Camber The small, upward curvature of a structural member as a compensation for deflection under load achieved, for instance, by pre-cambering glulam members and prestressing concrete members. Also the curvature of a road surface for drainage and safe cornering.

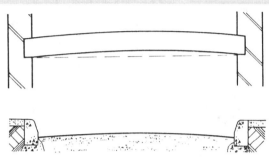

Camber

Camber arch A flat arch. It has a level extrados and a very slight rise in the intrados so that is does not appear to sag or deflect. *See also* Soldier arch.

Camber beam A beam that is cambered on its upper edge to compensate for deflection under load. *See also* Camber.

Camberboard A template used for setting out a camber.

Cant To tilt; to cut the wane from a log; a moulding formed from flat surfaces with no curves.

Cant bay A three-sided bay window with its outer sides splayed to the main wall.

Cant brick A splayed brick.

Canted A tilted, bevelled, splayed or off-square component or element.

Cantilever A projecting structural member such as a beam, slab, stair or truss that is supported at one end only.

Cap A block used to terminate posts. It covers the end grain in order to provide either a decorative finish or a weathering.

Capillary groove *See* Anti-capillary groove under Materials and scientific principles.

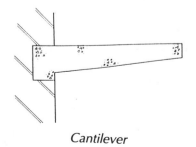

Cantilever

Capping   The top finish to a dwarf wall; a coping stone; the metal strip that covers a wood roll in flexible sheet metal roofing; also sealing the end of a pipe with a cap.

Carcass or carcase   The load bearing elements of a building e.g. walls, floors and roof. Also the main components of cabinets, standards, potboards and shelves etc.

Carcassing   The process of building the carcass. May also be used to describe the fixing of pipework and ducts for gas, water, electricity and ventilation. *See also* First fixing and Second fixing.

Carriage piece   The raking timber fixed centrally under wide stairs to provide additional support for the treads and risers. Brackets are fixed on either side of the carriage to support the full width of the treads. Also termed a rough string. Also used in formwork for concrete stairs, where it is placed on top of the flight and the brackets prevent the risers bulging due to concrete pressure.

Casement   The opening part of a casement window. *See also* Fanlight.

Casement door   A French casement.

Casement window   An opening window that is either top or side hung on hinges. Can be divided into traditional and stormproof depending on construction details.

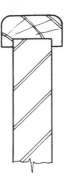

Capping

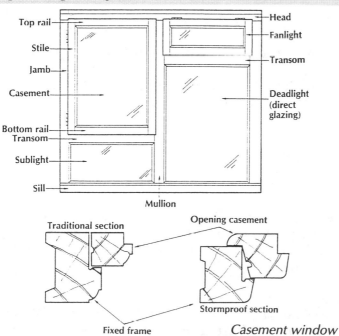

Casement window

Casing   The boxing-in of a pipe; a door lining; an apron lining.

Caulking   The sealing of a joint.

Cavity   A 50 mm to 75 mm air or insulated space between the internal and external leaves of a wall. *See also* Cavity wall. (85)

Cavity barrier   A piece of material which is used to close the cavity of a wall. It is positioned at intervals both vertically and horizontally in order to prevent the movement of smoke and flames within the cavity. *See also* Fire stop. (97)

Cavity fill   Thermal insulating material used to fill the cavity of a cavity wall, to reduce heat transfer.

Cavity flashing   A cavity tray.

Cavity tray   A DPC that is built into a cavity above openings. This will slope towards the outer leaf in order to direct away from the inner leaf any moisture in the cavity. *See also* Weep holes.

Cavity wall   A wall consisting of two leaves separated by a 50 mm to 75 mm cavity. The cavity is designed to prevent the transfer of moisture from the outside to inside leaf. Heat transfer through the wall can be reduced by filling the cavity with a thermal insulating material.

C/C   Centre to centre – refers to the spacing of members.

Ceiling   The upper horizontal surface of a room.

Ceiling joist   A joist that supports a ceiling but not a floor. Normally the ceiling at roof level but also those of a suspended ceiling.

Cement rendering   The plastering of an external wall surface with a cement and sand mortar. Normally to provide a weather resistant finish.

Cement screed   A layer of cement and sand mortar approximately 50 mm thick laid on the oversite of solid ground floors. It provides a smooth surface to receive the final floor finish.

Centre   The temporary support structure required when building an arch. *See also* Turning piece and Built-up centre. Also used to mean the middle of a component or object.

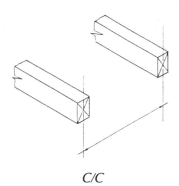

C/C

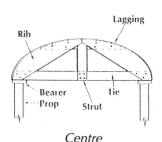

Rib
Lagging
Bearer
Prop
Tie
Strut

Centre

Chamfer A right-angled corner at the meeting of two surfaces that has been removed symmetrically thus at forty-five degrees, unlike a bevel which is asymmetrical. *See also* Cant, Splay and Tilted.

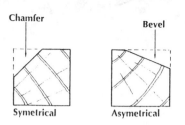

Chamfer

Chase A groove cut or built into the surface of an element, often to accommodate pipework for services.

Chase mortise A mortise that has a tapering chase alongside it. This enables a framing member to be inserted between the two other members which have been previously fixed e.g. the insertion of a new mullion between an existing window head and sill.

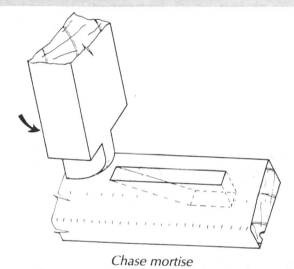

Chase mortise

Check A rebate.

Cill A sill.

Cladding The non-load bearing skin or covering of external walls, for weathering purposes. *See also* Brick skin, Shiplap boarding, Tile hanging and rendering.

Clapping stile A closing stile.

Clear span *See* Span.

Cleat A small piece of timber or metal fixed to one member and used to reinforce, positively locate or support another member.

Clench nailing Driving a nail so that it protrudes through timber, then bending over the point in the direction of the grain. Used in matchboarded door construction.

Close boarded Timber boarding fixed with their edges abutting tight without any spaces in between. (105)

Close-couple roof A form of couple roof where the feet of the rafters are closed with a tie, enabling an increased span. The closing tie also acts as a ceiling joist.

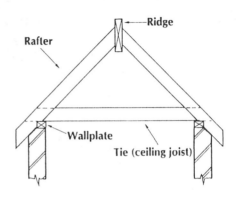

Close-couple roof

Close string A parallel stair string, the treads and risers being housed in its face and wedged in position. *See also* Cut string.

Closed eaves Overhanging eaves of a roof that have been closed with a soffit board rather than left open. *See also* Flush eaves.

Closed stair A stair built between two walls. Also termed a cottage stair.

Closer A brick cut along its length to reduce its width. Used next to the quion to close up the bond. *See also* King closer and Queen closer. (72)

Closing stile The stile of a door or window farthest from the hanging stile. Normally carries the lock or cockspur. (94)

Cockspur The item of ironmongery used to secure a casement window in the closed position. Also termed a casement fastener.

Collar-tie roof A form of close-couple roof where the tie is moved up the rafter to increase span or headroom.

Commode step A step with a curved tread and riser normally positioned at the bottom of a flight of stairs. *See also* Bull nose step and Curtail step.

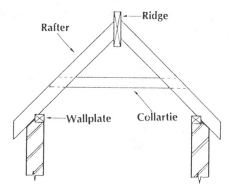

Collar-tie roof

Common rafters The rafters in a pitched roof that are at right angles to the wall plate and span from wall plate to ridge. *See also* Hip, Valley, Jack and Cripple rafters.

Component The various parts or materials that go together to form the elements of a building.

Composite construction Different materials used in conjunction such as a flitched beam formed from a steel plate sandwiched between timber.

Compound beam A timber beam built up from a number of pieces either nailed, glued or bolted with connectors. *See also* Composite construction, Flitched beam and Glulam.

Compound wall A wall constructed in two or more skins using different materials e.g. a timber-frame wall with brick veneer. *See also* Composite construction.

Concentric load A load that acts through the centre of a structural member as opposed to an eccentric one which does not.

Construction joint A joint formed to allow a break in concreting operations. Subsequently fresh concrete is placed against the joint which when hardened will be expected to provide structural continuity. *See also* Movement joint.

Continuous handrail The handrail of a geometrical stair.

Continuous string The string to a geometric stair.

Contraction joint A movement joint.

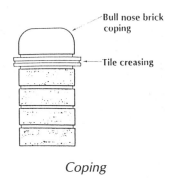

Coping

Cope   The laying of a coping. Also the cutting of, or scribing one moulding over another.

Coping   A protection to the top of an exposed wall e.g. a parapet using brick on edge, concrete, stone or metal. *See also* Creasing. (74)

Corbel   A brick, concrete, stone or metal component built into a wall surface and projecting from it in order to support a beam, floor joist or other load. *See also* Corbelling.

Corbelling   Brick or stonework in which each course projects progressively more past the face. *See also* Corbel.

Core rail   A steel rail that connects the tops of balusters of some stairs. The handrail, often geometric, being grooved out and fixed through it.

Cottage roof   A couple roof.

Cottage stair   A closed stair.

Counter battens   Two layers of timber battens fixed at right angles to each other, sometimes used for roofing and grounds for panelling. *See also* Brandering.

Counter ceiling   A false ceiling.

Counter cramp   An arrangement of three mortised timber battens fixed to the underside of timber counter tops over heading joints. Folding wedges are driven through the offset mortises to draw tight the joint.

Counter floor   A sub floor.

Couple-close roof   A close-couple roof.

Couple roof   A pitched roof consisting of pairs of rafters fixed at one end to the wall and at the other to the ridge. Spans are limited as the forces acting on the roof tend to spread the walls. *See also* Close-couple roof.

Coursed   Bricks, blocks, stone, slates and tiles laid in courses.

Courses   The parallel layers or rows of bricks, blocks, stone, slates or tiles including their bedding material.

Coursing joint   A bed joint.

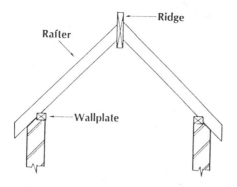

Couple roof

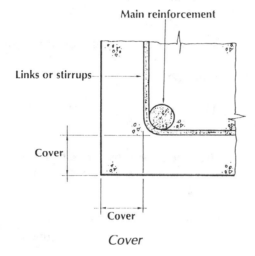

Cover

Cover The measurement between the face of a concrete element or item and its steel reinforcement. Cover is required to provide weather protection and fire resistance.

Cover fillet or cover strip or cover mould A narrow piece of trim used to cover the joints in wall panels, ceiling boards and abutting elements.

Cradling The rough framework or bracketing around a steel beam to support and provide a fixing for the covering material.

Creasing One or more projecting courses of tiles laid under a brick on edge, coping or at a sill. (82)

Cripple A member shortened in length e.g. a cripple stud is the shortened stud which carries the lintel over an opening in a timber-framed wall; a cripple rafter is the shortened rafter which runs from the ridge to the valley. *See also* Roof and Jack rafter.

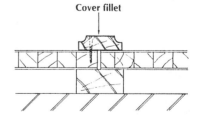

Cover fillet

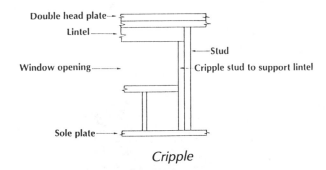

Double head plate
Lintel
Window opening
Stud
Cripple stud to support lintel
Sole plate

Cripple

Cross tongue   A narrow strip of plywood or timber section with diagonal grain. Glued into plough grooves to locate and join timber members. Also termed feather or loose tongue. *See also* Tongue and groove.

Crosscut   To saw timber at right angles to its grain.

Crown post   A king post.

Crown rafter   The central common rafter of a hipped end roof.

Curtail step   The half rounded or scroll end step at the bottom of a flight of stairs. It projects past the newel in two directions. *See also* Bull nose step and Commode step.

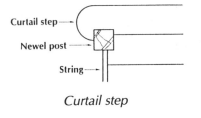

Curtail step
Newel post
String

Curtail step

Curtain wall   A non-load bearing infill wall unit used as the external enclosing wall for skeleton frame and cross wall construction.

Curtilage   The total extent of land occupied by, and attached to, a dwelling house.

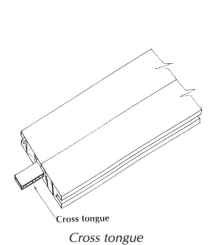

Cross tongue

Cross tongue

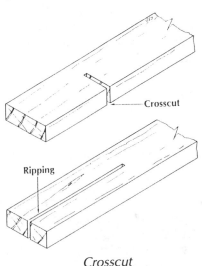

Crosscut

Ripping

Crosscut

Cut roof   A traditional pitched roof where the angles are cut on site as opposed to the use of factory-made trussed rafters.

Cut string   A stair string with an upper edge which has been cut to conform with the tread and riser profile. *See also* Bracketed stair.

Cutting list   A list of the materials required to make a building component or to carry out a particular task. A cutting list for a piece of joinery will itemize each part, give the number required and state their finished sizes.

Damp proof course   An impervious layer built into a wall about 150 mm above ground level, in order to prevent the rise of moisture by capillarity. Commonly termed a DPC. *See also* Damp proof membrane.

Damp proof membrane   An impervious layer incorporated into solid ground floors in order to prevent the rise of moisture by capillarity. Commonly termed a DPM. *See also* Damp proof course.

Dancing step   A winder or tapered step in a flight of stairs having its narrow end only slightly smaller than the fliers. Easier to negotiate than standard, sharply tapering winders but requires more space.

Datum   A known or assumed point, position or peg which is used to set out other levels or lines. *See also* Temporary bench mark and Ordnance datum.

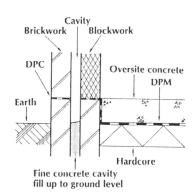

Damp proof course/membrane

Daylight size   The actual opening size of a window or glazed door through which light may travel. Also called a sight size.

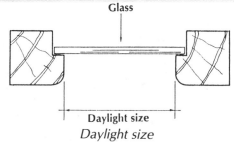

Daylight size

Dead shore   *See* Vertical shore.

Deadlight   A fixed section of glazing, it cannot be opened. *See also* Opening light. (77)

Deadload   The self weight of all the building materials used in the construction. Also the service installations and any built-in fitments. *See also* Imposed load and Load bearing.

Deadlock   A door lock which provides a straightforward, key-operated locking action; may be either rim or mortise type. Used for doors requiring security e.g. on a front entrance door in addition to a cylinder lock.

Deck or decking   The horizontal formwork and lining required for the in situ casting of concrete slabs; prefabricated units used for floor and flat roof construction. Also the boarding that supports the weathering or finishing layer of flat roofs and suspended floors.

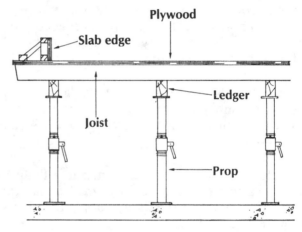

Plywood

Slab edge

Ledger

Joist

Prop

Decking

Deeping   The rip sawing of timber to the required thickness (cutting through the deepest section). *See also* Flatting.

Demolition   The dismantling or breaking up and removal of existing structures normally carried out prior to redevelopment work.

Diminished stile   A door stile that is narrower above the middle rail than it is below. Also termed a gunstock stile because of its appearance. Its purpose is to provide the maximum area of glass for daylighting purposes.

Discharging arch   A relieving arch.

Discontinuous construction   A break in the continuity of an element of construction, in order to minimize the transfer of impact structural bourne sound. *See also* Double partition and Floating floor.

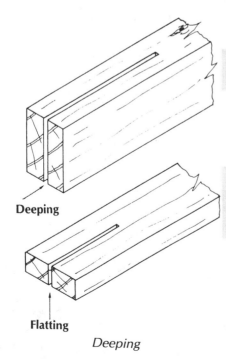

Deeping

Flatting

Deeping

Dog-legged stair A stair with two flights between storeys, connected by a rectangular half space landing. The two outer strings are tenoned into a common newel post thus there is no well. The stair takes its name because of its side elevational appearance.

Dog-tooth course A decorative projecting course of headers laid diagonally across a wall, so that only one corner of each brick projects.

Door A movable barrier used to cover an opening in a structure. Its main function is to allow access into a building and passage between the interior spaces. Other functional requirements may include weather protection, fire resistance, sound and thermal insulation, security, privacy, ease of operation and durability. May be classified by their method of construction: panelled; glazed; flush; matchboarded; fire resistant etc. and their method of operation: swinging; sliding; folding or revolving.

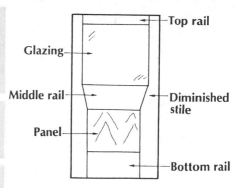

Diminished stile

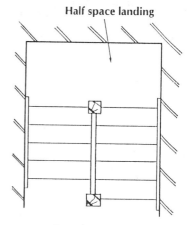

Dog-legged stair

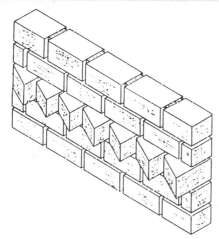

Dog-tooth course

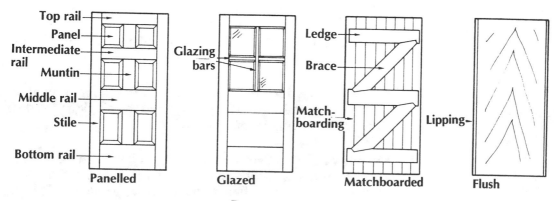

Door

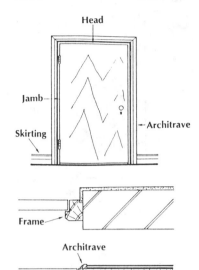

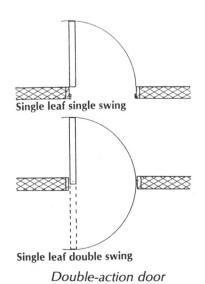

Door frame/door lining

Single leaf single swing

Single leaf double swing

Double-action door

Door blank An oversize, solid core, flush door manufactured so that it can be cut down on site to the required opening size. *See also* Blank door.

Door casing The trim (architrave) around a door opening.

Door frame The surround on which an external door or internal door is hung consisting of two jambs, a head and sometimes a threshold and transom. Normally with stuck-on solid stops and of a bigger section than door linings.

Door furniture The items of ironmongery required for a door's operation and security.

Door lining The surround on which mainly internal doors are hung, normally of a thinner section than door frames and often having planted stops. The main difference between door frames and door linings is that linings cover the full width of the reveal in which they are fixed, from wall surface to wall surface whereas frames do not. (66)

Door set A prefabricated assembly consisting of a door, hung in its frame or lining including stops, architraves, and furniture.

Door sill A threshold.

Door stop The strip of wood that is planted or stuck to the head or jambs of door frames or linings, it prevents the door from passing right through. Also a small block fixed to the wall or floor which prevents a door from opening too far causing damage. (66)

Dormer A vertical window that projects from the slope of a pitched roof and is normally provided with its own roof. *See also* Internal dormer and Roof light.

Dormer cheek The vertical, triangular side of a dormer window. Often clad with sheet metal.

Double action door A door that is able to swing through its opening in both directions, as opposed to a single action door which can swing in one direction only.

Double doors A pair of doors hung in one opening; may be single or double action. Double external doors with rebated meeting stiles are often termed french windows. *See also* Double margin door.

Double faced door A door with a different face detail on either side to match the decoration of the room or area in which each side faces. Normally constructed as two thin doors fixed back to back.

Double floor A suspended timber upper floor where the bridging joists are supported in mid span by a binder. *See also* Single and Framed floors.

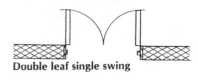

Double leaf single swing

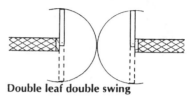

Double leaf double swing

Double doors

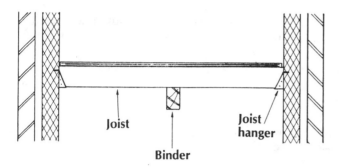

Joist   Joist/hanger

Binder

Double floor

Double glazing Glazing consisting of two panes of glass separated by an air space, used for thermal or sound insulation. Sealed units with typically a 25 mm air space are used for thermal insulation and secondary units with an air space of between 100 mm and 200 mm for sound insulation.

Double hung sash A sash window which slides vertically.

Double margin door A single door that has the appearance of a pair. Traditionally used for openings too narrow for a normal pair of doors, yet too wide for a standard single which would look out of proportion.

Double partition A partition wall consisting of two independent leaves with an insulated cavity between them. Forms a discontinuous construction used for sound insulation.

Double pitched roof A roof having two opposing pitched surfaces joining at the ridge, rather than a single or monopitched roof. Also termed a duopitched roof. *See also* Double roof.

Double roof A pitched roof that incorporates one or more purlins to provide intermediate support for the rafters, as opposed to a single roof having rafters spanning from wall plate to ridge without purlins. Also a term applied to a mansard roof. *See also* Double pitched roof.

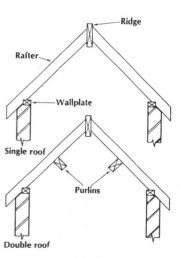

Ridge

Rafter

Wallplate

Single roof

Purlins

Double roof

Double roof

Double tenon   Two tenons arranged within the thickness of a member side by side. Used for thick members to increase the glue line and also traditionally for the middle rail of framed doors which were to receive a mortise lock, thus preventing the middle section between the normal twin tenons being cut away resulting in a loss of strength, where they are termed double twin tenons. (157)

Doubling piece   A tilting fillet.

Dovetail cramp   As a dovetail key but made from slate or metal and used to join stonework.

Dovetail key   A timber key used in joinery to connect timbers end to end, shaped like a butterfly or two dovetail pins joined together at their narrow ends.

Dowel   A short timber or steel circular rod used for fixing purposes. *See also* Draw-bore pinning.

DPC   A damp proof course.

DPM   A damp proof membrane.

Dragon piece   A horizontal piece of timber into which the hip rafter of traditional hipped end roofs was framed. It bisected the angle of the wall plates and was carried at its inner end by a dragon tie. *See also* Dragon beam under Architectural style.

Dragon tie   A tie spanning diagonally across the wall plates at an external angle, also supports the end of a dragon piece.

Draught bead   A deep staff bead at the sill of a sash window used to enable ventilation at the meeting rails of a partly opened sash while preventing draughts at the sill.

Draw-bore pinning   The drilling of a hole through a mortise and tenon joint for the insertion of a dowel. The hole in the tenon is located slightly nearer the shoulder so that the joint is pulled up tight when a tapered steel draw pin or pointed hardwood dowel is tapped in. Traditionally used on exterior joinery to secure joints. Also in common use for joints where cramping is inappropriate such as newel posts to stair strings.

Drip

Drip

Drip   A groove in the underside of an overhanging member, such as that in a window sill which is designed to stop water from running back under the sill to the face of the building. *See also* Throat. (157)

Dripping eaves   The eaves of a roof without a gutter.

Drop ceiling   A false ceiling.

Dry construction   Buildings which are constructed as far as possible without the use of wet trades. Maximum use being made of prefabricated elements and components. *See also* Timber-frame construction, Dry lining and Dry stone walling.

Dry lining   The finishing of internal wall surfaces with a wallboard, often the ivory surface of plasterboard with joints taped and filled, thus requiring no plastering.

Dry stone walling   Stone walls built without mortar also termed dyke walling.

Duct   A vertical or horizontal casing, chase, shaft or subway that accommodates pipes or cables in a building. Also the round or rectangular tubes used for ventilation and air conditioning systems.

Dumb waiter   A small non-pedestrian lift used to carry food and crockery from one level to another. May be manually or electrically operated.

Duopitched roof   A double pitched roof.

Dutch door   A stable door.

Dutchman   A piece of wood used to pack out badly cut joints. Also a piece of trim used to cover a mistake.

Dwang   Strutting between the joists of suspended floors.

Dwarf partition   A low partition wall that does not reach the full storey height.

Dwarf wall   A low wall that does not reach the full storey height, such as a sleeper wall that supports the floor joists of suspended ground floors.

Dyke   A dry stone wall.

E

Earth   The ground in general. Particularly excavated topsoil. Also an electrical connection to an earth electrode.

Easing   The process of planing the closing edges of doors and windows that are too tight within their frames; the shaping of a curved member to avoid an abrupt change in profile. Also the initial slackening of the folding wedges to shoring and arch centre supports after hardening to allow the structure to take up its load gradually.

Eaves   The lowest part of a pitched roof slope or the edge of a flat roof. Both usually overhang the wall and are finished with a fascia board. *See also* Verge. (104)

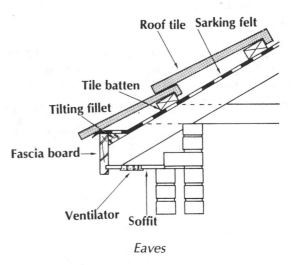

Eaves

Eccentric load   A load which does not act through the centre of a structural member, unlike a concentric load which does.

Edging strip   A lipping.

Effective span   *See* Span.

Elbow board   A window board.

Element   A constructional part of the sub or superstructure having its own functional requirements, such as foundations, walls, floors, roofs, stairs and structural framework. May be divided into primary, secondary and finishing. *See also* Component.

English bond   A brick bond for solid rather than cavity walls consisting of alternate courses of headers and stretchers. *See also* English garden wall bond and Stretcher bond.

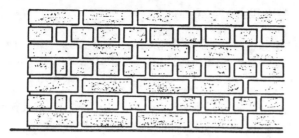

English bond

English garden wall bond A brickwork bond consisting of all stretchers except for a header course every five to seven courses. Also termed American bond. *See also* English bond and Flemish garden wall bond.

Erection The assembly, positioning and fixing of building elements. Mainly applied to factory-produced structural elements such as steel and pre-cast concrete items, timber-frame housing and also formwork. *See also* Striking.

Escalator A moving stairway with steps arranged on an endless belt.

Expansion joint A movement joint.

Extrados The upper edge of an arch or arch bricks or stones, as opposed to intrados which is the inner edge or soffit.

Eyebrow An eyelike dormer window in a pitched roof covered by an upward sweep of the main roof.

F

Fabric A building's carcass.

Face The surface of a building material with the best appearance; the widest surface of a material; also the front of a wall or building.

Face putty The triangular fillet of putty or other glazing compound holding the glass in position. Normally on the outside of a window. *See also* Back putty, Glazing and Glazing bead. (67)

Face side or edge The side or edge of a piece of timber that has the best appearance. The face is also the first to be prepared and serves as a datum from which the piece may be brought to width and thickness.

Facing The process of preparing a material's surface, also materials used to line or face a less decorative material.

Fair face A neatly built face brick wall; an internal brick or block which will be left unplastered; also a plain concrete high quality surface that has been produced from the form without any making good.

False ceiling A suspended ceiling below the structural soffit.

False tenon A loose or separate tenon inserted into the end of a rail; often to replace a damaged joint or in curved rails where the cross grain of a solid tenon would be likely to shear.

Falsework The temporary support structure that supports the formface of formwork. Although the term is mainly used to describe the temporary support structure for major civil engineering works such as large span bridges etc.

Fan truss A trussed rafter constructed using both vertical and inclined struts. *See also* Fink truss.

Fanlight A small window or light above a transom. Originally above an entrance door and having radiating glazing bars resembling a woman's fan. Also termed vent light or night vent when it can be opened. (77)

Fascia A deep board fixed to the end of rafters at the eaves of a roof. Also any other board set on edge and fixed to the wall face, such as a name board over a shop front. (92)

Feather A cross or loose tongue used to join wide boards edge to edge.

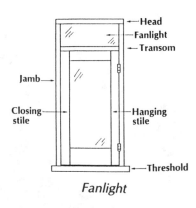

Fanlight

Feather edge A tapered section board used for close boarded fences and cladding; also applied to coping stones with its upper surface sloping or weathered in one direction only.

Fence The means of enclosing land to define boundaries and provide security.

Fender A baulk timber used in front of hoardings, scaffolds and temporary footpaths to protect users from traffic.

Fender wall A dwarf wall built under a suspended ground floor in front of a fireplace to support and contain the hearth slab.

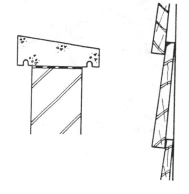

Feather edge

Fettle Any fine finishing work.

FFL Finished floor level.

Field The central portion of a fielded panel.

Fielded panel A panel having either its field raised above the margins, or its margins sunk below the field, or a combination of the two. *See also* Raised panel.

Fillet A small cross-section timber mould used to mask or cover joints. Also a triangular cross-section strip of mortar to weather an angle joint. *See also* Flaunching and Haunching.

Finger joint A lengthening joint for timbers, normally machine-made consisting of interlocking tapering fingers used in both joinery and structural work.

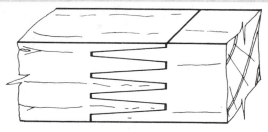

Finger joint

Finial An ornamental post, often pointed at the top of a gable or the apex of a turret roof into which the barge boards or hip ribs are tenoned.

Finished floor level (FFL) The actual final level or position of the finished floor including any screed, tiles or blocks etc., as opposed to the level of the concrete subfloor surface or floor joists.

Fan truss

Fink truss

Fink truss

Finishing element   The final surface of an element that may be a self finish such as fair face concrete and face brick or an applied finish such as plaster and paint.

Fink truss   A trussed rafter constructed so that all struts are inclined. *See also* Fan truss.

Fire back   The wall behind a fire.

Fire check door   A fire resisting door where its stated period of integrity is less than its period of stability e.g. a one hour fire check door might have sixty minutes stability but only forty-five minutes integrity. This is also known as a 60/45 fire resisting door.

Fire escape   A door or stair required to facilitate the escape of a building's occupants in the event of a fire. These are not normally used for everyday access.

Fire resisting door   A door having a stated period of stability and integrity e.g. a one hour door having stability of sixty minutes and an integrity of sixty minutes. This is known as a 60/60 door. Doors with an integrity less than their stability are commonly termed fire check doors.

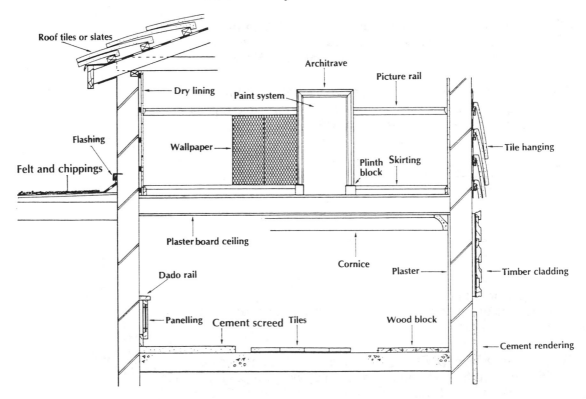

Finishing elements

Fire stop A strip of non-combustible material used in timber-frame construction to separate the cavities of semi-detached and terraced housing from the external wall or roof cavity. This along with cavity barriers serves to restrict the passage of smoke and flames.

Firring The long wedge shape battens fixed to flat roof joists before the decking, in order to provide a fall to the eaves.

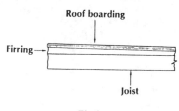

Firring

First fixing The fixing or installation of carpentry and joinery, flooring, frames, linings, studwork, window boards, plumbing, pipework, tanks, electrical, conduit, pattress boxes and wiring. *See also* Carcassing and Second fixing.

First floor The floor immediately above ground floor level.

First storey The space enclosed between the first floor and the roof or subsequent floor of a building.

Fitment Any fixed furniture such as kitchen units, wardrobes and other cupboard and shelf units.

Fixed-in Joinery that is inserted after the main building process, rather than built-in which is inserted during construction.

Fixed light A dead light.

Fixture Any item that is fixed to a building such as kitchen fitments, sanitary and light fittings etc. The removal of these would damage the property and thus should be sold with it.

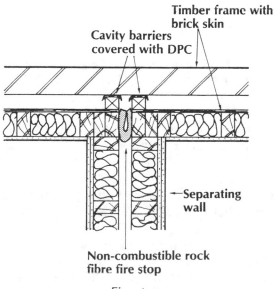

Fire stop

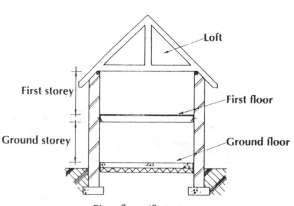

First floor/first storey

Flank wall   Any side wall of a building. *See also* Gable wall and Piend wall.

Flanking window   A window adjacent to an external door and sharing a common sill line. *See also* Side light and Wing light.

Flash   The fixing of flashings to make a weatherproof joint.

Flat arch   A camber arch. An arch with a level or almost level soffit. *See also* Soldier arch.

Flat joint   A flush joint.

Flat roof   Any roof having an angle or slope of less than ten degrees to the horizontal.

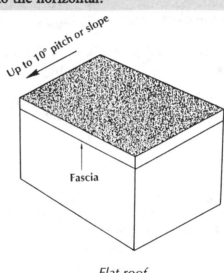

Flat roof

Flat sawn   Timber converted using the through and through method which produces about two-thirds tangential and one-third radial timber.

Flatting   The rip sawing of timber to the required width (cutting through the flattest section). The opposite of deeping. (86)

Flaunching   The sloping cement mortar fillet that surrounds and beds a chimney pot to its stack. *See also* Haunching.

Flaunching

Flemish bond   A solid rather than a cavity wall brick bond consisting of alternate stretchers and headers in the same course. *See also* Flemish garden wall bond and Stretcher bond.

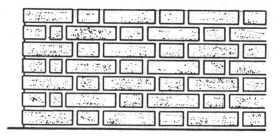

Flemish bond

Flemish garden wall bond A brickwork bond for solid walls consisting of three stretchers and one header. Also termed Sussex garden wall bond. *See also* Flemish bond and English garden wall bond.

Flier A step in a straight flight of stairs having a rectangular tread, not a winder or commode.

Flight A continuous set of steps between floors and/or intermediate landing levels. *See also* Stairway.

Flint wall A wall built using flints showing their napped faces, with brick or stonework quions.

Flitch The timber from which veneers are cut, a stack of cut veneers. Also a large section timber intended for reconversion.

Flitched beam A composite beam formed by sandwiching a steel plate between two timber beams and bolting together at intervals.

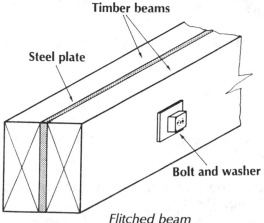

Flitched beam

Floating floor A discontinuous floor construction for impact sound insulation purposes whereby the load bearing structure is separated from the floor's upper surface.

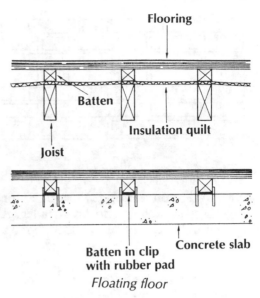

Floating floor

Floor The ground or upper levels in a building, which provides an acceptable surface for walking, living and working. *See also* Floating floor.

Floor joists A series of timber beams used in suspended floors to span the gap between walls and provide a flat fixing surface for the flooring. Variously named according to their size and function. *See also* Bridging, Trimmed, Trimmer and Trimming joists.

Floor strutting Herringbone or solid strutting to stiffen joists at their mid span.

Floorboards The close boarding fixed to the floor joists to provide a surface; normally either timber, plywood or chipboard.

Flush A surface in one plane. *See also* heading under Services and finishes.

Flush door A door having flush surfaces on both sides. It is not panelled, glazed or matchboarded. (87)

Flush eaves The eaves of a roof that do not overhang the fascia board.

Flush joint A brick or stonework mortar joint that is finished flush. *See also* Jointing.

Flush panel A panel in an item of joinery that finishes flush with its surrounding framework. May also be applied to a low level WC cistern. *See also* Bead butt.

Flying shore A horizontal shore.

Folded floor Floorboards that are sprung in place rather than cramped up.

Folding doors A door with two or more leaves hinged together so that they fold to one side, only the end leaf being hinged to the frame. May also have a top or bottom track. Often termed folding partitions as they are used to divide space.

Folding partition *See* Folding doors.

Folding wedges A pair of timber wedges used with their slopes opposed for levelling, easing and striking purposes in formwork and temporary supports.

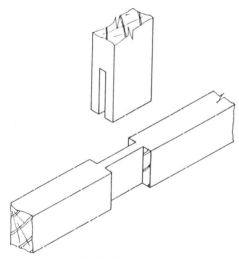

Forked tenon

Foot cut A seat cut. The horizontal cut of a birdsmouth at the foot of the rafter.

Footing The foundation to a wall.

Forked tenon A tenon formed in the centre of a rail to receive a muntin with a slot mortise.

Formwork A structure, usually temporary but can be partly or wholly permanent, which is designed to contain fresh fluid concrete, form it into the required shape and dimensions and support it until it cures sufficiently to become self-supporting. The surface in contact with the concrete is known as the formface while the supporting structure is sometimes referred to as falsework.

Formwork

Foundation The part of a structure that transfers the dead and imposed loads safely onto the ground. The most common types are strip, pad, pile and raft.

Foundation stone A large, inscribed commemorative stone built into the face of a wall close to ground level.

Foxtail wedging Secret wedging where the mortise is a dovetail shape.

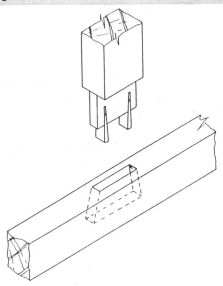

Foxtail wedging

Frame An assembly of components to form an item of joinery, such as a door or window; a structural framework of columns and beams or panels in steel reinforced concrete or timber. *See also* Skeleton frame, Platform frame and Balloon frame.

Framed door A door having a framework consisting of at least two stiles, top and bottom rail, jointed together with mortise and tenon joints or dowels. May also have intermediate rails, muntins, and panels. *See also* Frieze rail, Lock rail and Raised panel. (87)

Framed floor A suspended upper floor consisting of bridging joists that are supported in mid span by one or more binders, which themselves are supported by beams. Also termed triple floors, now mainly obsolete.

Framed grounds Grounds framed up, normally using either mortise and tenon joints or halvings. Used to provide a straight and level fixing surface for panelling. *See also* Counter battening.

Framed, ledged, braced and matchboarded door   A ledged and braced door with the addition of stiles and a top rail. The matchboarding is tongued into the top rail but passes over the middle and bottom ledges, which are themselves jointed into the stiles using barefaced tenons.

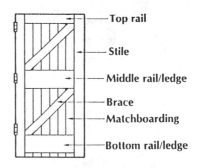

Framed ledged braced and matchboarded door

Framed partition   A stud partition that has been joined together prior to installation.

Franking   *See* Sash haunch.

Free standing   An item of furniture, joinery or other member that is not fixed to a wall.

French casement or french window or french door   External double doors leading to a garden, patio or balcony etc.

Frieze   *See* heading under Architectural style.

Frieze panel   The panel or panels above the frieze rail in doors and panelling.

Frieze rail   The uppermost intermediate rail of a door or panelling.

Front putty   The face putty.

Frontage   The boundary of a site that is parallel to the adjacent roadway.

Frontage line   The building line.

G

Gable The triangular portion of the end wall of a building with a pitched roof.

Gable coping The coping of a gable which projects above the roof line.

Gable end The end wall of a building having a gable.

Gable roof A double pitched roof with one or more gables.

Gable shoulder The projection of a gable springer from the face of the building.

Gable springer The projecting stone at the foot of the gable coping below the lowest kneeler. Its projection from the building's face is known as the gable shoulder.

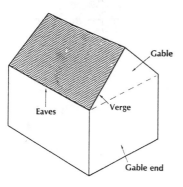

Gable roof

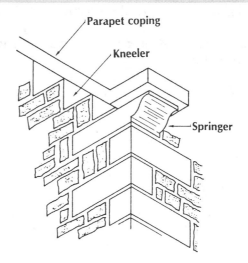

Gable springer

Gablet A small gable in a gambrel roof or over a dormer window.

Gambrel roof A double-pitched roof having a small gable or a gablet at the ridge level and a half hip below. *See also* Jerkin head roof.

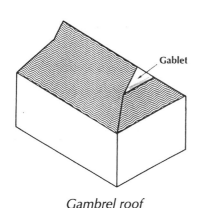

Gambrel roof

Gauge The proportions of different materials in a mix or the process of mixing two or more proportioned materials together; also the exposed face of a roofing tile or slate that is seen after fixing. This is also the centre for fixing the battens.

Gauged arch A brick arch built using rubber bricks with very fine joints.

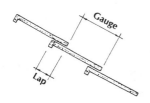

Gauge

Geometric stair A staircase normally without newel, where the outer curved string and handrail is continuous from top to bottom. May be circular or elliptical in plan. *See also* Helical stair.

Gin wheel or jinnie wheel A single pulley block and rope used to raise or lower loads e.g. buckets of mortar from ground to roof level. Also termed a jenny.

Glazing The process of installing glass into lights (doors and windows).

Glazing bead A small sectioned piece of timber or metal used to hold glazing in place.

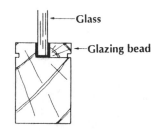

Glazing bead

Glue block A short, triangular section timber block glued to the underside of timber stairs to reinforce the joint between tread and riser or tread and string.

Glulam Glue laminated timber, normally structural timber members glued and layered up from fairly small sections and lengths to form either large cross-sections, long lengths, curved shapes or a combination of these.

Going The horizontal measurement of a step from the face of one riser to the face of the next. Stairs without risers are measured from nosing to nosing. The total going is the horizontal measurement of a stair flight from the face of the bottom riser to the face of the top riser. *See also* Going rod.

Going rod A length of timber on which is marked the individual step goings and total going for a flight of stairs.

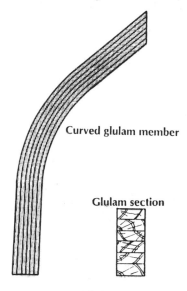

Glulam

Gravel board A horizontal board fixed between posts at ground level of good quality fences. Its purpose is to prevent moisture absorbtion into the end grain of the vertical pales, pickets or feather-edge boarding, which would rapidly lead to decay.

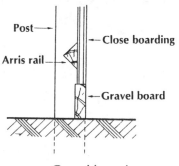

Gravel board

Green Freshly poured or layed concrete or cement mortar that has initially set but has not yet sufficiently hardened. Typically lasts for about seven days. So named due to its green appearance at this stage. Also timber with a moisture content over 18 per cent.

Grillage A built-up arrangement of timbers placed at right angles to the sole plate at the base of raking shores in order to transfer the load over a large area.

Grille An open grating or screen used for ventilation purposes.

Groin *See* heading under Architectural style.

Groove A continuous recess formed centrally along the edge of a timber board in order to accommodate a tongue. *See also* Tongue and groove, Cross tongue and Rebate.

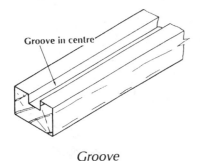

Groove in centre

Groove

Ground A surface which is to be painted, plastered or tiled. Also earth or an electrical earth connection.

Ground floor The floor of a building nearest the exterior ground level. May be solid or suspended/hollow.

Ground storey The part of a building between the ground and first floors. (97)

Ground work Site work that takes place in the ground such as excavations, drain laying, laying of paths and drives, etc.

Grounds Battens used to provide a straight and level fixing surface for panelling and skirting. *See also* Framed grounds, Counter battens and Soldiers.

Gunstock stile A diminished stile. (87)

Gusset A plate, either timber or metal, that is applied over abutting members to reinforce the joint e.g. nail plates in trussed rafters.

Half bat   A half brick in length. *See also* Half header. (72)

Half brick wall   A wall normally built in stretcher bond and used in cavity walls, having a thickness equal to half a brick's length. That is, the thickness of the wall is the width of a brick. *See also* Whole brick wall.

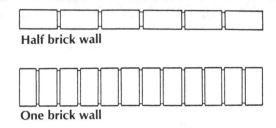

Half brick wall

One brick wall

Half brick wall

Half header   A queen closer. (72)

Half landing or half space landing   A landing joining two flights that turn 180 degrees such as a dog leg stair, its length being equal to the width of both flights. *See also* Quarter space landing.

Half lap joint   A halved joint.

Half pitch roof   A double pitched roof in which the rise is half the span e.g. forty-five degree pitch. *See also* Half span roof.

Half span roof   A lean-to or monopitch roof. *See also* Half pitch roof.

Halved joint   A timber lengthening or angle joint formed by cutting away half the thickness of each piece and fitting together flush. Used for wall plates and grounds etc.

Hammer beam roof   *See* heading under Architectural style.

Hammer headed key   A false tenon traditionally used to end joint curved joinery members. Both ends are shaped like hammer heads and fit into corresponding shaped slot mortices. Wedges are driven under the hammer head to pull the joint tight. *See also* Hammer headed tenon and Dovetail key.

Hammer headed tenon   A tenon in the shape of a hammer head formed onto the end of a jamb. It fits into a corresponding shaped mortise in the head. Traditionally used in curved head door and window frames. *See also* Hammer headed key.

Hand A term applied to doors and windows, determined by the hanging stile when viewed from the hinge knuckle side e.g. a left-hand door has its hinges on the left when viewed from the knuckle side.

Hand rail A horizontal or sloping rail at about waist height fixed to a wall or balustrade in order to provide a hand hold for users of the stairs. *See also* Hand rail scroll.

Hand rail scroll A spiral end to a hand rail.

Handed A description of hand e.g. right or left-handed. A handed pair refers to one being right-handed and the other left-handed. A handed pair of semidetached houses would have mirror image floor plans.

Hang The process of shooting in a door or window and fixing it to its frame or lining with hinges.

Hanging jamb or post The jamb or post of a door or gate on which the hinges are screwed. Also termed the hingeing jamb or post as opposed to the closing jamb.

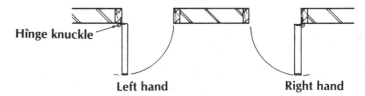

Hinge knuckle

Left hand   Right hand

Hand

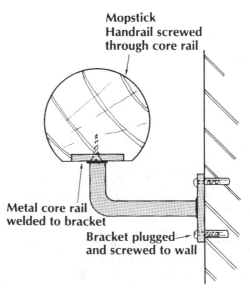

Mopstick
Handrail screwed
through core rail

Metal core rail
welded to bracket

Bracket plugged
and screwed to wall

Hand rail

Hanging stile   The stile of a door or window on which the hinges are screwed. As opposed to the closing or lock stile. (94)

Haunch   A shortened stub portion of a haunched tenon that has been reduced in width, either to ensure that its width does not exceed five times its thickness or to retain sufficient strength and enable wedging of a joint occuring at the end of a framework.

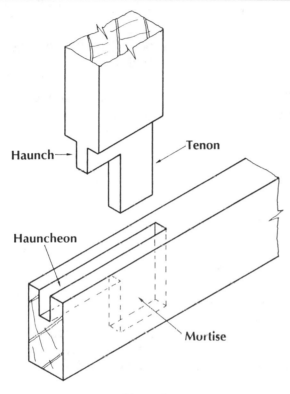

Haunch

Haunched tenon   A tenon that has been reduced in width to form a haunch. *See also* Twin and Double tenons.

Hauncheon   The stub mortise sinking adjacent to a through mortise which receives the haunch of a haunched tenon.

Haunching   The cement mortar fillet used instead of a flashing for weathering the joint between the chimney stack or projecting wall and pitch roof; the concrete surround to underground drainage pipes; also the stub mortise to receive a haunch.

Head   The top member of a frame or partition that is jointed to the jambs or studs. Also the bigger end of nails, screws and bolts. (77, 84)

Header A brick laid widthways across a wall so that it exposes its short face on the face of a wall. Also the exposed short face of a brick. *See also* Header bond and Stretcher.

Header bond and heading bond A brickwork bond where all bricks are headers, particularly used for curved plan walls.

Heading joint An end to end butt joint used as a lengthening joint for floor boards, matching and cladding etc. *See also* Splayed heading joint.

Headroom The vertical distance from the nosing line of a stair flight to the ceiling or bulkhead above.

Heel The end of a beam or rafter at its supports. Also the lower portion of a hanging stile. *See also* Toe.

Height board or rod A storey rod.

Helical stair A circular plan stair in which all the treads are winders. Also termed a winding stair. More commonly called a spiral stair.

Herringbone pattern A pattern formed by laying components at a slope in different directions to each other in alternate rows; used for decorative brick infilling, paving and woodblock floors. *See also* Parquet.

Herringbone strutting Diagonal cross strutting fixed across floor joists at their mid span to stiffen the joists and prevent lateral movement. *See also* Solid strutting. (154)

Hip rafter A pitched roof rafter used where two sloping roof surfaces meet at an external angle, providing a fixing point for the jack rafter and transferring their load to the wallplate. *See also* Valley rafter.

Hip rib A curved top hip rafter used for domed roofs. *See also* Jack rib.

Hipped end roof A double-pitched roof where the roof slope is returned around the shorter sides of the building to form a sloping triangular end. *See also* Gable end.

Hoarding A screen or fence, normally of a temporary nature, erected around a building site, demolition site or wasteland for the public's protection and safety, security of the valuables on the site and general aesthetics.

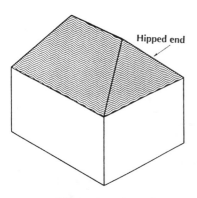

Hipped end roof

Hipped end

Hoggin   Coarsely graded fine aggregate or fine ballast used as an infilling or blinding material.

Hogging   A convex or cambered surface. *See also* Hollow.

Hogsback   A curved ridge tile which is parabolic in cross-section.

Holderbat   A metal fixing collar which clamps around pipes. A stand off leg fixes to the wall or soffit and keeps the pipe clear of the surface.

Holding down bolt   A bolt cast into a concrete element in order to provide a later fixing point. Often used to secure timber or steel-framed buildings to their foundations.

Hollow   A concave or sunken surface. Also a construction with an air space between two skins. *See also* Hogging.

Hollow core door   A flush door having air spaces between its two outer faces e.g. constructed with either a skeleton or cellular core.

Hollow floor   A suspended timber ground floor. Also a pot floor.

Hollow partition   A partition built in two separate leaves with an air space between forming a discontinuous construction for sound insulation. Term also used for partitions built using hollow bricks or pots.

Hollow wall   A cavity wall.

Honeycomb wall   A sleeper wall built in stretcher bond with a gap between bricks to permit air circulation under suspended timber ground floors.

Hood   A canopy or small roof over a door or window opening to throw off rainwater.

Hoop iron   Thin strips of iron or steel sometimes used to reinforce the bed joints of brickwork.

Hopper light or window   An inward opening, bottom hung casement window also known as a hospital window. *See also* Vent light.

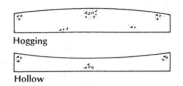

Hogging

Hollow

Hogging/hollow

Horizontal shore   Also termed a flying shore. Consists basically of a horizontal strut fixed at an upper floor level between two walls and suitably braced with rakers from above and below. Mainly used to provide temporary support when an intermediate building of a terrace has to be demolished prior to rebuilding.

Horn   The projection of the mortised member of a timber framework beyond the mortise. It provides strength when wedging and protection prior to use. May be cut off on hanging, built-in as a fixing, or left as a decorative feature. Also termed a joggle.

Hospital door   A flush door.

Hospital window   One or more hopper lights.

Housed joint   A sinking or recess in one piece of timber to receive another piece of timber or item of ironmongery.

Housed string   A close string as opposed to a cut string. Where a housing is cut to receive the tread and risers.

Housing   A housed joint. Also a quantity of dwellings.

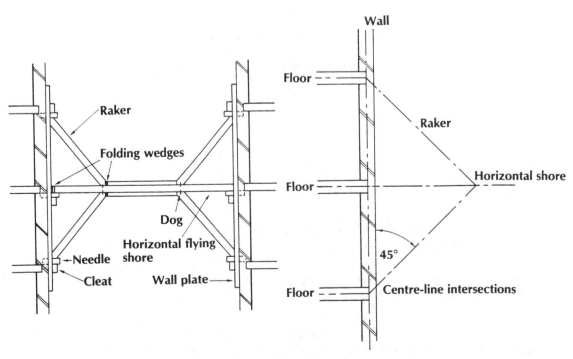

Horizontal 'flying' shore

Hyperbolic paraboloid roof   A form of shell roof construction created by raising the two diagonally opposite corners of a square to a higher level than the other two corners. A convex curve is produced between the low corners and a concave curve between the high ones.

I

Imposed load   The weight of any movable load in a building or structure, such as occupants, their furniture and other belongings. Also includes any environmental loads exerted e.g. wind, rain and snow. *See also* Load bearing and Dead load.

Impregnated   Timber that has been pressure treated with preservatives.

Improvement line   A line of intended improvement or widening to a road. These are established by a local authority and have a bearing on the building line.

In situ   In situation or in position. A term applied to building elements and components that are assembled or cast in their permanent position on site rather than prefabricated elsewhere e.g. in situ cast concrete rather than precast concrete.

In wind   Also winding a surface piece of material or joinery item that is twisted or warped. The opposite of out of wind.

Inclined shore   A raking shore.

Indent   A recess or groove below the main surface of an element or component.

Indenting   Toothing out.

Industrialized building   The prefabrication of building elements in factory conditions to minimize the amount of work carried out on site.

Infilling   Material used to fill the space in a framework. Also the process of adding additional buildings between existing ones.

Insert   A patch of veneer used to repair defects in the face veneer of plywoods, often boat shaped.

Insulate   To reduce sound or heat transfer through an element of construction normally by the inclusion of lightweight porous material, dense material or discontinuous construction.

Internal dormer   A vertical window within the general line not projecting above a pitched roof. *See also* Dormer.

Internal glazing   Glazing to internal lights in walls and doors etc. Also used to describe external glazing placed from within the building.

Intrados   The soffit of an arch or vault. *See also* Extrados.

Ironmongery   An overall term to include all small metal components such as hinges, locks, handles, nails, screws, and other fixing devices etc. Also known as hardware. *See also* Door furniture.

Jack arch   A Welsh arch.

Jack rafter   A pitched roof rafter that spans from the wall plate to the hip rafter, like a common rafter that has had its top shortened. *See also* Cripple rafter.

Jack rib   A curve top jack rafter used for domed roofs. *See also* Hip rib.

Jamb   A fixed outer vertical component of a window or door frame. *See also* Mullion. (77)

Jenny   A gin wheel. Also a pair of calipers with odd legs (one facing internally and the other externally).

Jerkin head roof   A double pitched roof which is hipped from the ridge part way to the eaves, the remainder being gabled. *See also* Gambrel.

Jig   A device used to hold work or guide tools.

Jinnie wheel   A gin wheel.

Joggle   A horn in framed joinery; a stub tenon; various methods of reinforcing the joint in block work or masonry walls.

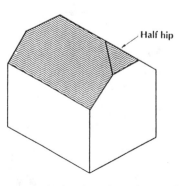

Jerkin head roof

Joinery   The making of timber building components such as doors, windows, stairs and cabinet construction etc., as opposed to carpentry.

Joint   The connection between building elements or components.

Jointing   The process of finishing the mortar joint of brickwork and blockwork as the work proceeds as opposed to raking out the joint and pointing at a later stage.

Joist   One of a series of parallel beams, often timber, that directly support a floor or ceiling surface. Variously named according to their function or position e.g. bridging, trimmer, trimming, trimmed, floor and ceiling etc. (86, 89)

Joist hanger   A metal strap used to support the end of a timber joist either to a wall or to a trimmer or trimming joist. *See also* Tusk tenon joint. (89)

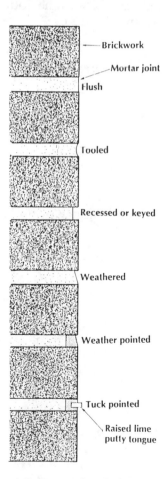

Brickwork
Mortar joint
Flush
Tooled
Recessed or keyed
Weathered
Weather pointed
Tuck pointed
Raised lime putty tongue

Jointing and pointing

K

Kerf   The width of a saw cut.

Kerfing   The technique of making partial saw kerfs in the back of a piece of timber to enable it to bend around a curve e.g. fitting skirting to a segmental bay window.

Key   The roughness of a surface which may provide a mechanical bond. Also an insert let into a joint to strengthen it.

Keyed   Held by a key.

Keyed joint   A joint held by a key. Also both the tooled and recessed jointing of brickwork.

Keyed mortise and tenon   A tusk tenon.

Keystone   The central voussoir of an arch.

Kick plate   A metal plate fixed to the bottom rail of a door to protect it from kicking during use. .

King closer   A closer brick cut diagonally from the centre of one header face to the centre of the stretcher face. *See also* Bond. (72)

King post   The vertical timber member of a king post truss.

King post truss   A traditional timber roof truss consisting of two principal rafters, a tie beam, a king post and two struts. *See also* Queen post truss.

Kite winder   The central winder of three turning a quarter space. *See also* Stairs.

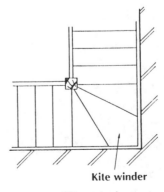

Kite winder

Kite winder

Knee   A handrail ramp which is convex on its upper surface.

Knee brace   A brace between a roof truss and a column or wall, used in some structures to increase resistance to wind load.

Kneelers   The level bedded, sloping topped stones of a gable end. *See also* Gable springer. (104)

Knocked down   A building element or component that is delivered to the site unassembled e.g. door linings and built-in fitments.

Knuckle   The part of a hinge containing the holes through which the pin passes. Also the joint between the two slopes on the side of a mansard roof.

L

Laced valley   A swept valley.

Lacing   Generally tying together for stability. In formwork the horizontal members used in conjunction with adjustable steel props to space and tie them together. Their use greatly increases the safe working load.

Lagging   The thermal insulation of pipes and tanks. Also the small sectioned timbers around arch centres used to support the walling material. (78)

Laminated timber   *See* Glulam.

Landing   A level platform at the end, or between, flights of stairs. *See also* Half space and Quarter space landing.

Lantern   A raised light on a flat roof often taking the form of a miniature greenhouse-type structure. *See also* Skylight and heading under Architectural style.

Lap joint   The joint between components which overlay each other.

Lapped tenons   Two tenons which enter a mortise from opposite ends and bypass each other within the joint.

Lashing   A fibre or steel rope for securing the top of a ladder to a scaffold platform.

Lath   A small section of timber about 5 mm × 25 mm, formally used as a base for plaster. *See also* Lath and Plaster under Services and finishes.

Lathing   Base to receive plaster. Formally timber laths but now used for plasterboard strips and expanded metal etc.

Lattice window   Any window with small diamond-shaped panes of glass, mainly leaded lights.

Lay bar   Any horizontal glazing bar.

Lay board   The board fixed to the rafters of a main pitched roof at the valley in order to receive the foot of the secondary roof's cripple rafters.

Lay light   A light positioned horizontally in a ceiling. *See also* Lantern and Skylight.

Lay panel   A panel having its greatest dimension horizontal.

Layer board   The board forming the base of a box gutter on which the lining is laid.

Leaded light   *See* heading under Architectural style.

Leaf   One of a set of doors or windows hung in the same opening. Also one of the two skins of a cavity wall.

Lean-to   A monopitch roof abutting a higher wall or building.

Ledge   One of the horizontal timbers on the back of a matchboarded door.

Ledged and matchboarded door   A door made from matchboarding, clench nailed to ledges. Diagonal braces may be added, forming a ledged braced and matchboarded door thus reducing the tendency to sag. *See also* Framed, ledged, braced and matchboarded door. (87)

Ledged, braced and matchboarded door   A door made up from matchboarding, clench nailed to ledges and diagonally braced to resist sagging. *See also* Framed ledged, braced and matchboarded door.

Ledgement   A string course.

Ledger   A timber or proprietary aluminium member which is used to support the decking joists of slab formwork. Also known as runners or primary beams. (86)

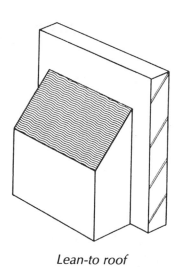

Lean-to roof

Let in Sunk flush with the surface. *See also* Sinking.

Light *See* heading under Architectural style.

Lightning conductor A thick, copper lead connected to earth and projecting above a building. It greatly reduces the chance of the building being struck by lightning.

Lining A door lining, the surround to a sash window reveal; sheet material used to cover wall surfaces. Also a paint finishing defect.

Lintel The small beam over doors and windows to span the opening and transfer the wall load to either side. (73, 84)

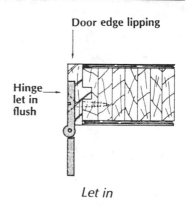

Let in

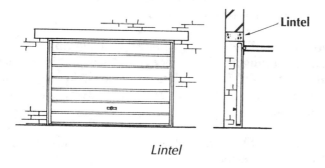

Lintel

Lipping The thin strip of timber fixed to the edge of flush doors in order to cover the edge of the facing. (87)

Load bearing A building element that carries or transfers any dead or imposed loads, as opposed to non-load bearing.

Loading coat A concrete floor laid over tanking in a basement floor in order to prevent it being displaced by upward water pressure.

Lock block A block of timber provided in hollow core flush doors in a suitable position to accommodate a mortise lock or latch and provide a fixing for the lock or latch furniture.

Lock rail The horizontal member of a door into which a lock can be mortised or fixed. Also known as a middle rail.

Lock stile The vertical member of a door which provides a fixing for a lock. Also known as the closing stile.

Loft The roof space under the slopes of a pitched roof between the rafters and uppermost ceiling joists. May be used for storage. *See also* Attic under Architectural style. (97)

Louvre

Loose tongue   *See* Cross tongue.

Louvre   A ventilator formed using horizontal sloping slats to exclude rainwater; made from timber or glass and may be fixed or pivoted.

Made-up ground   Ground which has been filled or raised in level using earth or hardcore.

Maintenance   The keeping, holding, sustaining or preserving of a building and its services at an acceptable standard to enable it to fulfil its function. This can take the form of a planned maintenance programme or emergency corrective maintenance.

Making good   The rectification of defects in a building or its services. Also termed snagging.

Mansard roof   A double-pitched roof where each slope has two pitches. The lower part has a steep pitch, the upper part rarely exceeds thirty degrees. The ends may be hipped or gabled. *See also* Knuckle.

Margin   The projection of a closed stair string above the tread and riser intersection; the gauge of a tile or slate; the border around the field of a panel or other member; also the face of stiles or rails.

Margin light   A narrow pane of glass at the edge of a sash window or door.

Knuckle

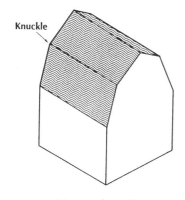

Mansard roof

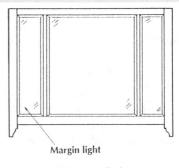

Margin light

Margin light

Mason's mitre A moulding joint in stone or timber that has the appearance of a mitre, but has in fact been carved out of the solid, the actual butt joint between the two members being set back from the corner.

Matchboard Tongued and grooved timber boards, often with a moulded edge, used for flooring, cladding, wall and ceiling lining etc. (87)

Matching Matchboard. Also used to describe the arrangement of timber veneers, such as book matching, slip matching and quarter matching.

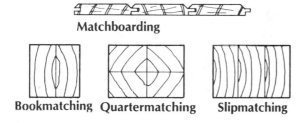

Matchboarding

Bookmatching Quartermatching Slipmatching

Matching

Mattwell A recess in the floor surface just inside the entrance door to accommodate a mat for wiping feet.

Meeting rails The rails of a sash window that come together when the window is closed.

Meeting stiles The stiles of double doors that come together when the doors are closed.

Meeting stiles

Member A component part of a building element or a sub part of a component itself.

Membrane A thin sheet of impervious material used to prevent the passage of moisture. *See also* Damp proof membrane.

Mezzanine *See* heading under Architectural style.

Mid feather A parting feather or wagtail. Also the leaf of brickwork which separates two flues in a stack.

Mismatching A bad fit at a joint, poor grain or colour matching of veneers. Also a step, wave or other deviation in a concrete surface.

Mitre A butt joint between two members at an angle to each other, the line of joint being a bisection of the angle e.g. forty-five degrees for a right-angled joint. A mitre between members of dissimilar section is called a bastard mitre.

Moisture barrier A waterproof barrier used under external cladding. It will prevent the passage of moisture but at the same time will allow water vapour to pass through, thus the construction is able to breath. *See also* Vapour barrier. (164)

Monk bond A modified Flemish bond consisting of two stretchers and one header repeated in each course.

Mopboard A skirting board.

Mopstick A circular cross-section handrail with a small flat surface on its underside to receive brackets or balusters. (108)

Mortise A rectangular slot cut in a timber component to receive a tenon cut in another, or a lock. In general the width of the mortise should not exceed one-third of the thickness of the timber. Also a similar slot cut in stonework to receive a fixing cramp or lewis for lifting.

Mortise and tenon A framing joint between timber components normally at right angles to each other e.g. the rails of a door which are tenoned into mortises cut in the stiles. (109)

Mosaic *See* heading under Architectural style.

Mould Formwork for a precast concrete component. Also used as a collective term for a range of fungi which grow on damp walls and ceilings.

Moulded A description applied to any material having a moulding.

Moulding A shaped piece of building material (often timber, stone or plaster) used either for decoration or as a weathering. In joinery mouldings are termed as either stuck or planted. Also the process of shaping a piece of material. *See also* heading under Architectural style.

Movement joint A joint incorporated in concrete to allow expansion or contraction from its original position. Exposed edges of the joint are filled with a waterproof sealing compound to prevent erosion and water penetration. *See also* Expansion, Contraction and Construction joints.

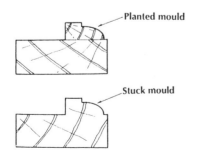

Planted mould

Stuck mould

Moulding

Mullion The intermediate vertical member of a window or door frame. It separates the openings or lights and is jointed into the head and sill. *See also* Muntin, Transom. (77)

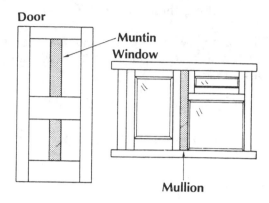

Mullion/muntin

Muntin The intermediate vertical member of a framed door or panelling. It separates the panels and is jointed into the rails. /*See also* Mullion.

Mushroom headed column The splayed shaping at the heads of some columns (normally found where slabs do not incorporate beams). These support the loads of large span floor slabs and transfer them onto the columns. Also creates a distinctive decorative feature.

Needle A short horizontal timber or steel beam inserted through a wall and supported by vertical shores used as temporary support when part of the wall below is to be removed. Also the member inserted in a wall to provide an abutment for raking shores. (112, 136)

Newel The large, sectioned, vertical member or post at either end of a flight into which the string is jointed. *See also* Newel drop. (84, 149)

Newel cap The decorative wooden top of a newel post.

Newel drop The projection of the upper newel of a flight below the ceiling level. Also termed a pendant newel.

Nib A small projection from a surface.

Night vent A ventlight or fanlight.

Nog or nogging A short horizontal piece, or pieces, of timber fixed between the vertical studs of a timber-framed wall or partition. Their use stiffens the studs and also provides intermediate fixing points for the covering material. *See also* Brick nogging. (154)

Nog brick/block A brick or block made from a nailable material. These are built into the wall to provide a convenient fixing point for joinery and trim. *See also* Brick nogging.

Non-load bearing The opposite of load bearing. A building element that does not, or is not intended to, carry any dead or imposed loadings.

Nosing The front overhanging edge of a stair tread. The finish to the floor boards around a stairwell opening or the top narrow tread of a timber flight where it adjoins a landing or upper floor. *See also* Apron lining and Nosing line. (149)

Nosing line An imaginary inclined line which would touch the upper edges of the nosings of a flight of stairs. Also known as the pitch line.

OBM   An ordnance bench mark.

Observation panel   A glazed panel mainly in flush doors to allow observation.

One brick wall   *See* Whole brick wall. (107)

Open eaves   The overhanging eaves of a roof which are not closed by a soffit. The underside of the rafters is therefore exposed.

Open floor   A suspended floor without a ceiling, thus having exposed joists. *See also* Open roof.

Open mortise   A mortise cut on the end of a component. It is open on three edges, used in bridle joints and with forked tenons. Also termed slot mortise.

Open newel stair   An open well stir.

Open plan   The design of buildings with the minimum of fixed internal partitions.

Open plan stair   Stairs with open risers.

Open roof   A roof without a ceiling, thus having its rafters exposed. *See also* Open floor.

Open stair   A stair which is freestanding or abuts a wall on one side only.

Open string   The outer string of a stair that faces, or is open to, the well. Can be either cut or closed. (149)

Open well stair   A half turn stair that incorporates two newels at landing level to separate the strings of the upper and lower flights. Thus a central open space or well is formed. Can be termed an open newel stair. *See also* Dog leg stairs.

Opening light   The portion of a sash or casement window that may be opened for ventilation rather than a dead light which is fixed.

Ordnance bench mark   A mark found cut into the wall of churches and public buildings, consisting of a broad arrow topped by a horizonal line. It is established by the Ordnance Survey and represents a level above the ordnance datum. Used for establishing levels related to construction works. *See also* Temporary bench marks.

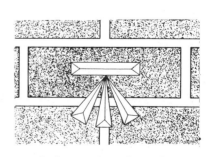

Ordnance bench mark

Ordnance datum The mean sea level at Newlyn in Cornwall which forms the reference level for ordnance bench marks.

Ordnance Survey A government survey of Great Britain which covers all land and construction, published as a series of maps which cover the entire country.

Oriel window A bay window that projects from an upper storey only.

Orientation The positioning of a building in relation to the points of a compass.

Out of wind A surface, piece of material or joinery item that is flat and true as opposed to in wind or winding.

Outer string An open string. (149)

Outside glazing Glazing that is installed from the exterior of a building as opposed to inside or internal glazing which is installed from the interior.

Overcloak The part of a member that overlaps the one below. Particularly applied to overlaps in sheet metal roofing. *See also* Undercloak.

Overhand work External brick walls laid from the inside of the building rather than by erecting an external scaffold.

Overhang The part of an element or component that projects past its support. *See also* Cantilever.

Overhanging eaves The eaves of a roof that project past the line of a building rather than flush eaves.

Overhead door An up and over door often used for garages. It is opened by raising and sliding into a horizontal track.

Oversailing course A string course or corbelling.

Oversite A layer of concrete under a ground floor.

Pad An isolated mass of concrete forming a foundation. Also a piece of timber built in a brick joint to provide a fixing point for frames etc.

Padstone A stone or concrete block built in a load bearing wall under heavy loaded points, such as the end of a beam or lintel, to spread its load.

Pair Two matching items; may be identical or handed.

Pale The narrow, vertical, pointed boards of a fence or palisade. Also termed picket.

Palisade A fence formed using pales.

Pallet A twisted section of timber used for fixings.

Pane A glass light. *See also* heading under Architectural style.

Panel The glass, timber, sheet material that is used as an infill between the framing members of doors and panelling: brick, stone or concrete infill between a structural framework. Also an area of concrete between supports. *See also* Bay.

Panelled door A framed door with panels between the framing members e.g. a raised and fielded panel door. (87)

Panelling Framing consisting of stiles, rails and muntins, mortised and tenoned together, with panels infilled between framing members. Used to line walls and sometimes ceilings. Term also loosely applied to wall linings made up of sheet material.

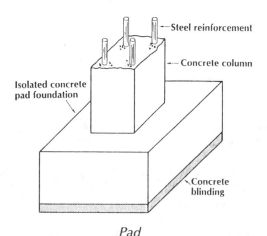

Pad

Paramount partition   A proprietary partitioning system consisting of plasterboard sheet sandwiched either side of a cardboard lattice core. These are fixed between the head and sole plates. Battens are inserted into the core where sheets abut.

Parapet   The projection of a wall above roof level. (104)

Parapet gutter   A concealed box gutter behind a parapet wall.

Parquet floor   A wood block floor finish, often layed in a herringbone pattern.

Parting bead   The narrow section of timber that is grooved into the pulley stiles and head of a sash windowframe to separate the upper and lower sashes. (73)

Parting feather   A mid feather or wagtail.

Parting wall   *See* Party wall.

Partition   An internal wall used to divide space, but may sometimes be load bearing. *See also* Stud and Paramount partitions.

Party wall or parting wall   A shared, separating wall between two dwellings.

Pavement   A footway for pedestrians, suitably prepared with either asphalt, bitumen, bricks, concrete or flags etc, to withstand any expected traffic and weathering.

Pavement light   Glass blocks set into a pavement to provide daylight for a basement below. Also termed a vault light.

Pavilion roof   A roof having a regular polygon plan shape and equal hips on all sides.

Pellet   A small, circular section, cross-grained piece of timber with matching grain and colour to that in which it is inlaid. Inserted into counterbored hole and pared flush with the surface to conceal fixing screws.

Pelmet   A vertical board or short curtain fixed over a window to conceal the curtain rail or blind fixings.

Pendant   A member that is suspended or projects below the main surface e.g. a pendant newel which is housed around a trimmer and terminates just below the ceiling line.

Pergola   *See* heading under Architectural style.

Permanent formwork   The formwork or formface that is not struck but is left permanently in position after the concrete has been cast.

Perpends   The vertical/perpendicular joints in brick and blockwork. Often abbreviated to perps.

Perps   Perpend joints.

Picket   A fence pale or palisade.

Pier   A thickening of a wall to provide stability. *See also* heading under Architectural style.

Piend   A hip or hip end wall.

Pilaster   A pier, especially beside doors and window openings. *See also* heading under Architectural style.

Pigeon hole wall   A wall built in honeycomb bond.

Pile   Supporting, concrete column-like members. In situ cast below ground level in compacted or drilled holes to act as the foundation for a structure. Can also be precast members which are driven or screwed into place.

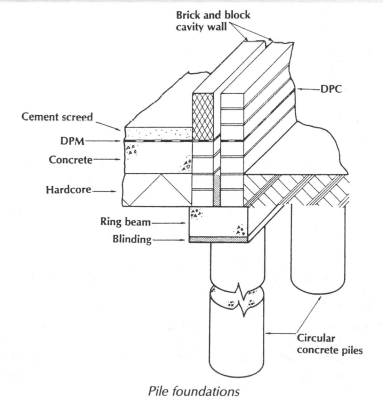

Pile foundations

Piling The process of forming a pile foundation.

Pillar A column or pier.

Pilotis *See* heading under Architectural style.

Pinch rod Two battens used in such a way that they slide apart to check dimensions between walls, floors and openings etc.

Pitch The angle of inclination to the horizontal of roofs and stairs; or the ratio of rise to span e.g. a one-third pitch roof having a span of 6 m will rise 2 m. Also a term used for the distance between items of uniform spacing e.g. reinforcement links in concrete spaced 150 mm apart, have a pitch of 150 mm.

Pitch line *See* Nosing line.

Pitched roof A roof having a sloping surface in excess of ten degrees pitch. Variously named according to their shape. *See also* Flat roof.

Planted A moulding bead or stop that has been fixed in place by nailing or screwing or with adhesive rather than being cut from the solid. *See also* Stuck. (122)

Plate A horizontal timber used to provide support and fixing point for rafters and studs. Also serves to spread their load uniformly along their supporting member e.g. wall plate, sole plate. *See also* Gusset.

Plate cut A foot or seat cut on a rafter at the wall plate. Part of a birdsmouth.

Platform frame A type of timber-frame construction in which the walls are constructed in single storey height panels; floors extend over the wall panels at each level thus forming a platform on which the walls above can be erected. *See also* Balloon frame.

Plinth The recessed base of a cupboard to provide foot space, also termed kick board. *See also* heading under Architectural style.

Plinth block A block of timber traditionally fixed at the base of an architrave to take the knocks and abrasions at floor level. Also used to ease the fixing problems at this point, which occur when skirtings are thicker than the architrave. (66)

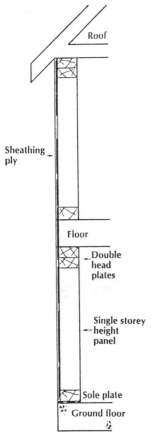

Platform frame

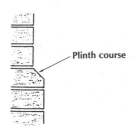

Plinth

Plinth course The projecting course or courses at the base of a wall, forming the plinth.

Plug tenon A stub tenon with four shouldered faces. Also termed a spur tenon.

Plumb Vertical.

Plumb cut The vertical cut of a rafter at the ridge. Also the vertical cut of a birdsmouth joint at the wall plate.

Pocket The opening in the pulley stile of a box frame sash window through which access to the weight is gained. An opening cut into a wall in order to receive a beam end. Also an opening cast in a concrete member to provide a later fixing point.

Pointing The process of filling a mortar joint after raking out and working it to the required joint profile. Carried out as a secondary operation after the wall has been built, unlike jointing which is carried out as the wall is built. *See also* Repointing. (115)

Popping Blowing of plasterwork from its backing, particularly over the nail heads used to fix plasterboard.

Portal frame A single storey structural frame consisting of a rigidly jointed vertical column and sloping roof beam. Mainly used in pairs joined at their apex.

Post A main vertical building support member, mainly in timber. *See also* Stud and Puncheon.

Pot floor A concrete floor slab in which hollow clay blocks known as pots have been cast in the bottom. Their use reduces the amount of concrete required and the slab loading.

Precast A prefabricated concrete item that is cast out of its final location, either on site or in the factory.

Prefabricated A building element, component or even an entire structure that has been largely constructed in a factory prior to assembly on site. *See also* Industrialized building.

Primary element The main supporting, enclosing or protection elements of a building. Also those that divide space and provide floor to floor access. *See also* Secondary and Finishing element and Components. (132)

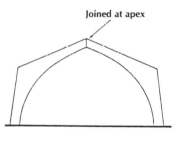

Joined at apex

Portal frame

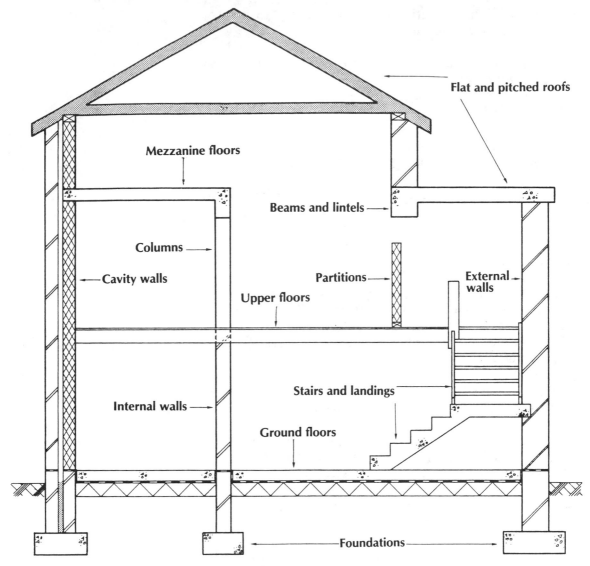

Flat and pitched roofs

Mezzanine floors

Beams and lintels

Columns

Cavity walls

Partitions

External walls

Upper floors

Internal walls

Stairs and landings

Ground floors

Foundations

Primary elements

Principal A traditional roof truss or a principal rafter. *See also* King and Queen post truss.

Principal rafter The rafter of a traditional roof truss that provides support for the purlin which in turn supports the common rafters.

Profile A horizontal board supported on pegs and set up just outside the proposed foundation trenches at all wall intersections. On top of the profile will be four nails or saw cuts to indicate the edges of the trench and the wall faces. Also the outline of a mould or a template used to reproduce or shape a mould.

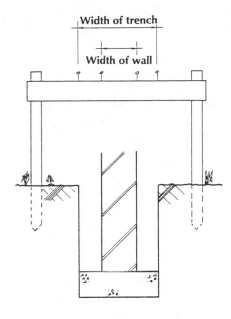

Profile

Prop A vertical, temporary support member used in form-work and shoring. Sometimes termed puncheon. *See also* Adjustable steel prop. (78, 86)

Protected shaft A stairway, lift shaft, escalator or service duct opening that permits the passage of items, and or persons, from one fire resisting building compartment to another.

Pugging A soundproofing material added to timber floors and partitions.

Pulley stile The vertical stile of a box frame sash window into which the pulleys are fixed. (73)

Punched work The face of ashlar, worked with a mason's punch, having rough, diagonal strokes across the surface.

Puncheon A vertical timber post. *See also* Stanchion.

Purlin A beam used in a double roof to provide support for the rafters at their mid span.

Purlin roof A double roof.

Push plate A finger plate.

Pyramid roof A pavilion roof.

Q

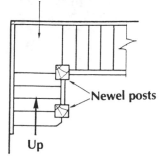

Quarter space landing

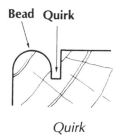

Quirk

Quadrant A curved metal casement stay for hopper windows; also used to describe a quarter round moulding.

Quarter round A quarter round moulding mainly termed a quadrant mould.

Quarter space landing A square landing having sides equal to the width of the stairs. Mainly used to turn a stair ninety degrees. *See also* Half space landing.

Queen closer A brick cut along its length to form a half header. Used next to the quion to close up the bond. *See also* King closer. (72)

Queen post The main vertical members of a queen post truss.

Queen post truss A traditional timber roof truss designed to support purlins. Similar to a king post truss except that it has two vertical queen posts positioned at one-third spans instead of the central king post.

Quetta bond A one and a half brick wall with both faces in Flemish bond. The resulting half brick square gaps are filled with concrete and steel reinforcement as the work proceeds. *See also* Rat trap bond.

Quick step Small radius circular moulding or work.

Quirk A recessed narrow groove alongside a mould.

Quirk bead A semicircular mould run on the edge of a piece and separated from the main face by a quirk.

Quoin The external corner of a building or the brick or stonework used to form the external corner.

Quoin header The corner header in a wall it will also be the quoin stretcher in the return wall.

Quoin stretcher The corner stretcher in a wall it will also be the quoin header in the return wall.

Racking back   The process of building up the corners or ends of a brick wall first. The remainder of the wall being stepped down or racked back towards the midde. *See also* Toothing out.

Racking back

Raft foundation   A foundation in the form of a reinforced concrete slab with thickened edges, covering the entire base area of a building.

Rafter   A sloping timber in a pitched roof, spanning from wall plate to ridge, variously named according to their position e.g. common, crown, hip, valley, principle. When cut short at either end they are termed jack or cripple. (70, 80)

Rail   A horizontal member of a timber framework variously named according to their position: top, bottom, middle, lock, intermediate and frieze. A handrail. Also a horizontal fence member.

Raised panel   A thick panel, normally in a door or wall panelling, where the edges are reduced in thickness to tongue into the framework.

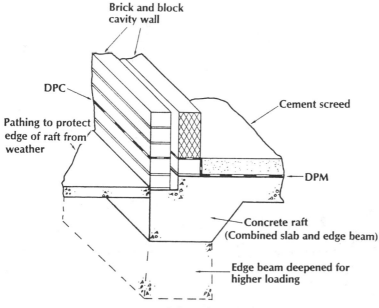

Raft foundation

Raised and fielded panel   A raised panel, having distinct raised margins and a central field. May also be raised, sunk and fielded.

Rake   An inclination to the vertical e.g. a raking shore, a raking riser.

Raking out   The cleaning out of a mortar joint to a depth of about 20 mm in preparation for pointing.

Raking riser   A stair riser that is inclined so that its top edge or nosing overhangs part of the step below.

Raking shore   A temporary support for a building consisting of a large section, inclined member carried at its base on a soleplate and located at the top by a wall piece, needle and cleat. *See also* Rider shore and Grillage.

Ramp   An inclined footpath, corridor or floor gently rising from one level to another without a vertical step. Also a bend in a handrail.

Random courses   Walling, slating or tiling having courses of varying depth.

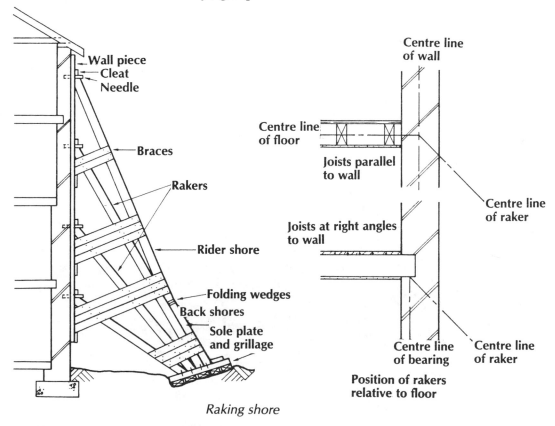

Raking shore

Random rubble *See* heading under Architectural style.

Rat trap bond A brick on edge bond used to build an economy one brick wall. Each course consists of alternate headers and stretchers resulting in a cavity between the stretchers similar to quetta bond, although they are normally left unfilled. Also termed a row lock bond.

Rebate A long rectangular recess cut at the edge of a timber member rather than a groove which is cut centrally along the edge e.g. the rebate around a window to receive the glass.

Recessed joint A mortar joint that is set back from the face of the bricks.

Rejointing Pointing brickwork.

Relieving arch An arch built above an opening which is also spanned by a lintel – so named because it relieves the lintel of some of its loading. Also termed a discharging arch.

Repointing The raking out of old mortar joints to a depth of about 20 mm, filling with fresh mortar and working to the required profile.

Retaining wall A wall which has a higher level of earth on one side than the other. It is designed to withstand overturning due to the pressure exerted by the higher ground.

Return The change in direction, normally right angles, of a wall or building component.

Reveal The vertical part of an opening in a wall that is not covered by a frame.

Reveal lining The finish to a reveal.

Revetment The facing of a sloping earth bank with stones, bricks or blocks to prevent slippage and erosion.

Revolving door A door normally having four leaves fixed at right angles to each other – opens by revolving on a central pivot. Often used as an entrance to public buildings in order to minimize heat loss and draughts.

Ribbon A ribbon course; a joist housed into the studs of a balloon timber-frame to provide a bearing for the ends of the bridging joists. Also, in formwork, a horizontal member fixed to decking which prevents beam sides and slab edges spreading. They may also be used to strut off. (67)

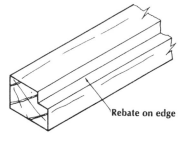

Rebate on edge

Rebate

Ribbon course An occasional course of decorative tiles and slates used in walling.

Rider shore The outer raker of a multiple raking shore that is supported on a back shore rather than continuing down to the sole plate and grillage. (136)

Ridge The apex of a double-pitched roof. Also the horizontal board at the apex which provides a bearing and fixing point for the rafters. (80, 89)

Ridge capping or tile The covering at the ridge of a roof which weathers the joint between the two slopes.

Rim latch A latch that is fixed to the face or rim of a door rather than one fitted into a mortise.

Rim lock A lock that is fixed to the face or rim of a door rather than one fitted into a mortise.

Rip To saw timber parallel to its grain, as opposed to crosscutting. *See also* Deeping and Flatting. (84)

Rise The vertical distance: from wall plate to ridge of a roof; from springing line to the highest intrados point of an arch; from the surface of one tread to the surface of the next in a stair. Also refers to the total rise of a flight of stairs which is measured from the lower finished floor or landing level to the next finished floor or landing level above.

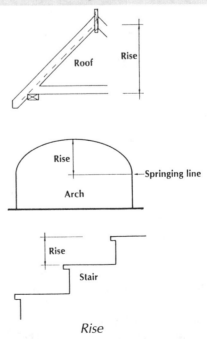

Rise

Riser The vertical member of a step. *See also* Tread. (149)

Rod A board, batten or drawing on which the dimensions of an element or component are set out full size. *See also* Setting out.

Roof The uppermost external envelope of a building that spans the walls. Mainly flat or pitched but variously named according to its shape.

Roof light A skylight.

Roof space A loft.

Roof truss An arrangement of principal rafters, posts and a tie which are designed to support a purlin e.g. king or queen post trusses. *See also* Trussed rafter.

Rough arch A relieving arch.

Rough axed Bricks, mainly for arches, that have been cut with an axe or bolster and not rubbed.

Rough string The carriage under a wide flight of timber stairs.

Rough work Walling that will later be covered with plaster, render, facing brick or cladding etc.

Roughing out The process of forming the approximate shape of an item before commencing the final, accurate shaping and finishing.

Rowlock A brick on edge course. Also applied to rat trap bond.

Rubbed arch An arch built using rubbed bricks. Also termed a gauged arch.

Rubbed brick A rubber.

Rubbed finish A concrete surface that has been finished by rubbing down with a carborundum stone.

Rubbed joint A glue joint formed between the edges of two boards which, after applying the adhesive, are rubbed together to expel air and excess adhesive.

Rubber A soft, frogless brick suitable for rubbing into shape for a gauged arch.

Rubble   Broken brick, concrete, stone and similar materials. May be used for hardcore.

Rubble walls   Stone walls built using stones that have not been finally worked such as random rubble walls.

Runner   A binder in floor and roof construction. Also a ledger in formwork.

Running bond   Stretcher bond.

Rustics   A facing brick with a textured surface formed by either impressing a pattern on it or facing it with sand.

S

Saddle back coping   A coping with a pointed apex formed by two weathered upper surfaces.

Saddle stone   A triangular-shaped stone used at the apex of a gable.

Safety arch   A relieving arch.

Sandwich beam   A flitched beam.

Sandwich construction   Composite construction.

Sarking   Roofing felt or timber boarding laid under the slates or tiles of a pitched roof. (92)

Sash haunch   A joint used for small, section-framing members, especially sash windows, where the sash cord groove would seriously weaken a normal haunched mortise and tenon joint. Instead of forming a normal haunch on the rails tenon and a haunching adjacent to the stiles mortise these locations are reversed. Also termed a franking.

Sash window   A window which slides open horizontally or vertically rather than swinging or pivoting. *See also* Box frame. (166)

Scarf   A lengthening joint for timber and veneers; the ends are cut at a splay and glued together. Structural members may also be bolted, spliced or wedged.

Score   The scratching of a surface to improve the mechanical bond.

Screed   A cement and sand mortar laid over concrete floors to level out any irregularities and provide a smooth finish for the final floor covering.

Scribe   The cutting of one member to fit another, such as cutting the bottom edge of a skirting board to fit an uneven floor surface. Also the marking of a surface using a pointed tool. Also used for skirtings at internal angles.

Scribed joint   A joint used between the stuck moulded edges of framed joinery, where the moulding of the horizontal member is scribed over the vertical one. Used in preference to a mitre as it masks the effect of any shrinkage.

Second fixing   Items in a building that are fixed after plastering – doors, trim, fitments, sanitary appliances, boilers, radiators and electrical fittings. *See also* First fixing.

Secondary element   The non-essential elements of a structure. These have a mainly completion role around openings in primary elements and within the building in general. *See also* Finishing elements and Components.

Secret dovetail   A dovetail formed within a mitre joint.

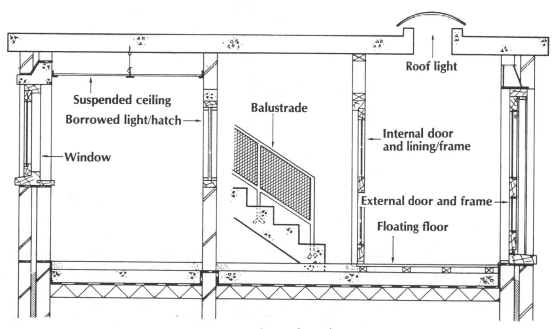

Secondary elements

Secret fixing Any method of fixing an item that cannot be seen on completion such as the secret nailing of floorboards.

Secret nailing Nailing that cannot be seen on the surface. Mainly carried out on tongue and groove boarding by nailing through the tongue and laying the next board to cover the nail heads.

Secret screwing Slot screwing.

Secret wedging A method of securing stub tenons which are inserted into blind mortises. Small wedges are driven into saw cuts made in the tenon's end. When the joint is cramped the wedge expands the tenon causing it to grip the sides of the mortise securely. Also known as foxtail wedging when the mortise is dovetail-shaped.

Services The provision of mechanical and electrical services in a building.

Setback The upper storey or storeys of some high-rise buildings that are progressively moved back from the building's face in order to ensure sufficient light in the road below.

Sett A small, rectangular block of stone, often granite and used for decorative paving.

Setting out The process of establishing pegs, profiles and levels for excavation and positioning buildings or marking out the positions of walls on a floor slab. Also drawing of full-size rods for the production of joinery items.

Sheathing Boarding or sheet material fixed to timber-framed walls and pitched roofs to provide bracing. Also the formface of vertical formwork such as walls. (164)

Shiplap boarding Timber cladding having a rebate on one edge which locates into a barefaced tongue on the other.

Shoddy Squared granite stones less than 300 mm thick. Also a slang term for any inferior quality work.

Shoot To straighten a board edge or door etc. with a shooting plane.

Shore A temporary support member which may be vertical (dead shore), horizontal (flying shore) or inclined (raking shore). *See also* Shoring.

Shoring A temporary support system consisting of shores used to give stability to a building while it is undergoing structural repairs or alterations.

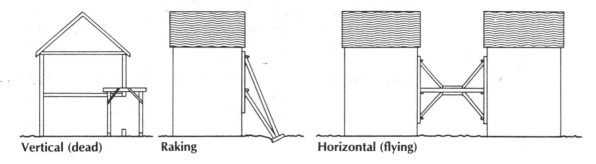

Vertical (dead) Raking Horizontal (flying)

Shoring

Shoulder The surfaces either side of a tenon at its root that abuts the timber either side of a mortise. *See also* Barefaced tenon.

Shutter A timber or metal cover that is fastened over windows and doors at night to provide security. Also an item of shuttering.

Shuttering Formwork for in situ cast concrete. *See also* Shutter.

Shutting jamb or post The jamb or post against which a window, door or gate closes.

Shutting stile The closing stile of a door or window.

Side light A glazed panel adjacent to a door. *See also* Wing light and flanking window.

Siding Cladding or weather boarding.

Sight size The actual opening of a window or glazed door which can be seen through. Also termed daylight size.

Sill The lower horizontal member of a fixed window or external doorframe. Also the projection below a window or door opening. Both are normally weathered on their top surface to provide a rainwater run off. Sometimes spelt cill. *See also* Sole plate, Subsill and Threshold. (77)

Sillboard A window board.

Silver lock bond Rat trap bond.

Single floor   A suspended timber upper floor where the bridging joists are supported at either end only. *See also* Double and Framed floors.

Single roof   A pitch roof without purlins, the common rafters of which span from wall plate to ridge. *See also* Monopitch roof. (89)

Sinking   A recess or groove cut below the main surface of a component, often to receive an item of ironmongery, such as the leaf of a hinge or the face plate of a lock.

Skeleton construction   A reinforced concrete or steel-framed structure consisting of columns and beams.

Skeleton core   Light section, internal framework of a hollow core flush door.

Skeleton stair   An open riser stair.

Skew   Out of square, any oblique angle.

Skewback   The sloping upper surfaces of a springer or the springer itself.

Skewnail   Nails inserted obliquely to the timbers being joined, alternate nails being driven in the opposite direction for increased resistance to withdrawal.

Skirting   The horizontal trim, often a timber board, that is fixed around the base of a wall to mask the joint between the wall and floor. Also protects the plaster surface from knocks at low level. *See also* Plinth block. (66)

Skylight   Glazing in a roof, either opening or dead, used for additional lighting or ventilation. Also termed roof light. *See also* Lantern.

Skyscraper   A high-rise building.

Slab   A concrete floor or roof. Also a paving slab. (86)

Slamming stile   The closing stile of a door or window to which the lock or handles are fixed, as opposed to the hanging stile. Also known as the clapping stile.

Slat   A thin strip of material e.g. a timber slat.

Slate boarding   The close boarding of a roof.

Sleeper   A large section timber used to spread the load and provide a fixing point or bearing, such as the half baulks used to provide a temporary foundation under site hutting etc. Also the wall plate on a sleeper wall.

Sleeper wall   A dwarf wall built on the oversite concrete to provide intermediate support for the joists of suspended timber ground floors. *See also* Honeycomb wall.

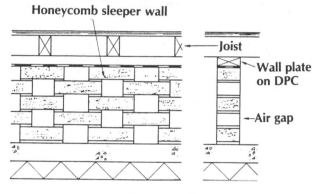

Sleeper wall

Slip   A narrow brick or a thin piece of timber.

Slip feather or tongue   A cross tongue.

Slip mortise   An open mortise. Also a chase mortise.

Slot mortise   An open mortise.

Slot screwing   Secret fixing of panelling and trim; projecting screw heads are inserted into keyhole shaped slots.

Snagging   *See* Making good.

Snap header   A half bat that appears on the face of a wall as a header.

Snow board   A horizontal board or wire guard fixed on the slope of a pitched roof just above the eaves' gutter to prevent snow from sliding and damaging the gutter or its fixings. Also a grid placed over a box gutter to prevent it filling with snow.

Soffit   The underside of a building element or feature surface e.g. a concrete slab, beam, stair or arch etc. In formwork the soffit formface is called decking.

Soffit board The board that closes the overhanging eaves or verge of a roof. (92)

Soldier Applied to short building components used vertically e.g. vertical bricks such as in a soldier arch or soldier course; vertical grounds for fixing skirtings; vertical noggins wedged between the flanges of a steel beam.

Soldier arch A flat or slightly cambered arch or lintel spanning an opening; formed using brick on end soldiers.

Soldier course A decorative string course incorporated in a brick wall; formed using brick on end soldiers, often of a contrasting colour to, or projecting from, the surface.

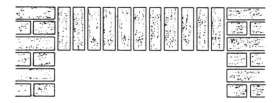

Soldier arch

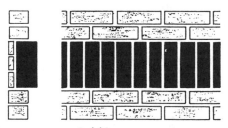

Soldier course

Sole plate A horizontal member, also known as a sill, which is placed under vertical supports to provide a fixing point and ensure even distribution of loads. *See also* Stud partition. (67, 84)

Solid core door A flush door having a solid rather than hollow core.

Solid floor A solid concrete ground floor that is constructed on the earth rather than suspended. Also a solid concrete suspended floor rather than one using hollow pots (pot floor).

Solid strutting Horizontal strutting between upper joists at mid span using solid timber rather than herringbone. (154)

Span The horizontal distance between the supports of a structural member (joist, lintel, beam, slab, roof etc.). Clear span is measured between the faces of supports while effective span is measured between the two centres of bearing.

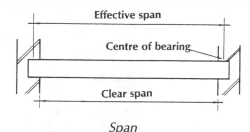

Span

Spandrel The triangular area formed under a flight of stairs or either side above the extrados of an arch. *See also* Spandrel step and heading under Architectural style.

Spandrel step A triangular cross-section stone or concrete step which, with others, forms a flush soffit. *See also* Spandrel.

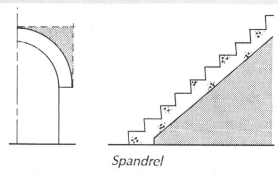

Spandrel

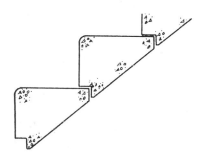

Spandrel step

Spar A rafter.

Spar piece A collar tie.

Special A component that has to be specially ordered or made.

Spiral stair A helical stair.

Splay A large bevel or chamfer that extends across the full width of a surface.

Splayed coping A coping with a weathered or splayed top face, giving it a wedged-shaped cross-section.

Splayed grounds Grounds with a splayed top edge used at skirting level. Also acts as a screed for the plaster.

Splayed heading joint   A floorboard heading joint cut out of vertical so that an overlap is formed between the ends of the two boards.

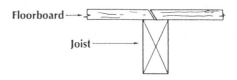

Splayed heading joint

Splice   A halved lengthening joint between two timbers which is covered on either side by a gusset. Also refers to the letting or sinking-in of a small piece to repair a damaged item of joinery.

Split course   A brick course built using bricks that have been cut lengthways in order to reduce the course's height. (72)

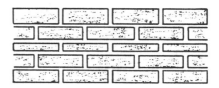

Split course

Springer   The first stone of an arch that is bedded on the springing line. Its sloping top surface is termed the skewback. *See also* Gable springer.

Springing line   The horizontal line in an arch where the intrados starts to move away from the vertical reveal e.g. a horizontal line at the start of the curve.

Springings   The point of intersection on either side of an arch between the intrados and wall on the springing line.

Sprocket   A wedged-shaped piece of timber or a separate short rafter applied to the common rafters at the eaves of a roof to slacken the pitch. This slows down the rainwater at this point and reduces the tendency for it to overshoot the gutter. Termed sprocketed eaves.

Sprocketed eaves   The eaves of a roof finished with sprockets. Also termed bellcast eaves.

Squint quoin   The quoin at the corner of a building that is not at right angles.

Stable door A door being divided horizontally in half, each being hung separately. Also termed a Dutch door.

Staff bead The bead around the inside of a sliding sash window that holds the inner sash in place. (73)

Stair or stairs The series of steps that form a stairway. Can be classified according to their plan shape e.g. straight flight, quarter turn, half turn, and geometrical or method of construction, e.g. close string, cut or open string, open riser, carriage beam and spine beam. (75)

Staircase Formally the space within which a flight of stairs was built, now used to mean the flight itself.

Stairway A series of steps giving floor-to-floor access including any balustrading or handrail. Each continuous set of steps in between floors or landings is termed a flight of stairs.

Stairwell The gap between two parallel flights of stairs. Also the well or space in which a stairway is built.

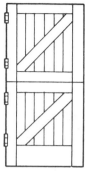

Stable door

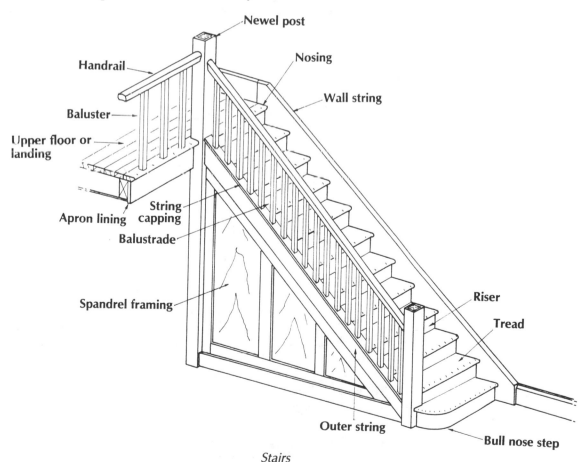

Stairs

Stall riser   The vertical surface below a shop window, stall-board and the pavement.

Stallboard   The sill and its framing below a shop window and over the stall riser.

Stallboard light   A pavement light adjacent to a stall board.

Stanchion   A column. Normally a steel universal section. *See also* Puncheon.

Steel casement   A casement window made of steel. Often incorporated into a timber subframe.

Step   The combination of a tread and riser. A series of steps form a stair flight or stairway.

Stepped flashing   See heading under Services and finishes.

Stepped foundations   A level foundation in sloping ground formed in a series of extended steps to avoid excessive excavation.

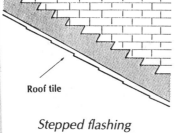

Roof tile

Stepped flashing

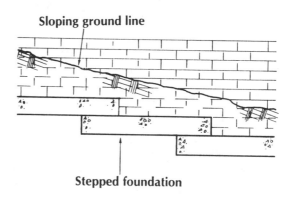

Sloping ground line

Stepped foundation

Stepped foundations

Sticker   A small section piece of timber used to separate adjacent lengths of timber when forming a stack. Their use prevents distortion of the timber and permits a free air circulation. *See also* Seasoning under Materials and scientific principles.

Sticking   The process of moulding a piece of timber. *See also* Stuck moulding.

Stile   The outer vertical member of a door or opening part of a window. *See also* Muntins and Mullions. (73, 77)

Stock Converted timber. Also any other material that is normally readily available from a supplier's stock. Hence stock bricks, stock sizes etc.

Stooling The built-up ends of a concrete or stone stooled windowsill, so formed to provide a flat surface on which to bed the walling at the jambs.

Stop *See* Door stop.

Stop moulding A moulding that stops part way along a length of material.

Stopped mortise A blind or stub mortise.

Storey The space between one floor level and the next floor level in a building.

Storey rod A length of batten used to measure a storey on which relevant positions may be marked e.g. brick courses, sill height, lintel height, joist hcight and the risers of a stair. Also termed a height board.

Stormproof window A form of high performance casement window construction designed to give increased protection from wind-driven rain. Normally incorporating double rebates and/or proprietary seals. (77)

Straight arch A camber or flat arch may be formed using soldiers.

Straight flight A stair flight that does not change direction between floor levels. All of its treads will be fliers.

Straight joint A butt joint in brickwork. A break in bond resulting in perps being in line one above the other.

Straining piece A member fixed on a horizontal shore to take the strain where the rakers abut.

Stress graded The classification of timber for strength may be carried out visually where it is based on the KAR or by machine which measures deflection under load.

Stressed skin panel A structural unit consisting of timber-framing members on which a plywood skin is fixed to one or both sides. The panel functions as a whole transmitting the stresses between the skin and framing. Used as prefabricated panels for floors, walls and roofs.

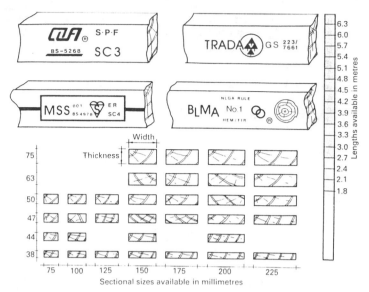

Stress graded

Stretcher A brick laid lengthways along a wall so that it exposes its long face on the face of a wall. Also the exposed long face of a brick. *See also* Stretcher bond and Header.

Stretcher bond A brickwork bond where all bricks are stretchers, except for the quoin header. Particularly used for the outer leaf of cavity walls.

Stretching course A course of stretchers.

Strike The process of removing a temporary support structure such as formwork to concrete, arch centres and shoring. *See also* Easing.

Striking wedges Folding wedges.

String The inclined board of a stair, also termed a stringer, into which the treads and risers are cut or housed. Variously named according to their position or type e.g. wall string, outer string, close string, cut string and wreathed string. (84)

String course

String course A decorative, often projecting, course of bricks around a building at either windowsill, head or upper floor level. A projecting string course will throw rainwater clear of the building. *See also* Soldier course.

Stringer A stair string.

Strip foundation A narrow strip of concrete in the bottom of a trench on which the wall is built. Termed a deep strip when filled to almost ground level.

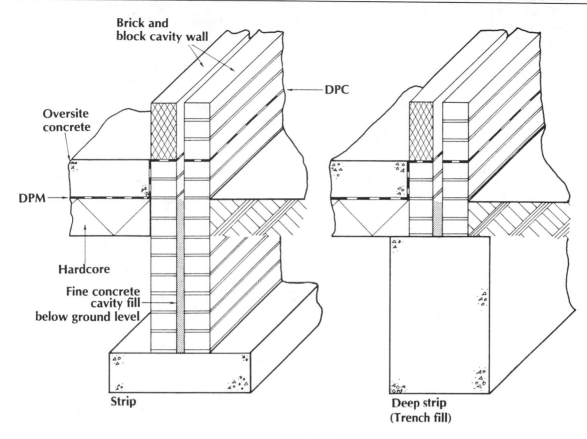

Strip foundations

Strong back A proprietary metal soldier used in concrete wall formwork where high pressures are anticipated.

Struck joint A weathered joint.

Structural Load bearing; any building element or component that carries or transfers a load in addition to its own weight.

Structure The load bearing or structural primary elements of a building; may be divided into substructure and superstructure depending on their location.

Strut A structural member that is mainly subjected to compression unlike a tie, which is mainly in tension. *See also* Brace. (78)

Strutting The use of struts such as herringbone or solid struts in flooring or diagonal strutting to hold vertical formwork in position.

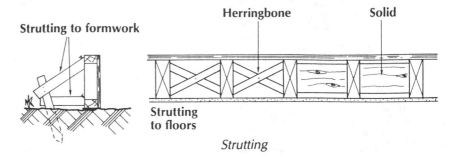

Strutting

Stub tenon   A tenon that fits into a blind mortise. It stops short and does not pass completely through the mortised member. *See also* Through tenon. (71)

Stuck moulding   A moulding that has been worked on the solid timber section rather than one which is planted. (122)

Stud   A vertical timber or metal member of a partition wall fixed between the sole plate and head plate.

Stud partition   A partition wall built of timber or metal studs, fixed between a sole and head plate, often incorporating noggins as stiffeners.

Subbasement   The storey or storeys below the basement.

Subfloor   A structural floor that carries the imposed loadings but does not provide the floor finish. *See also* Counter floor and Floating floor.

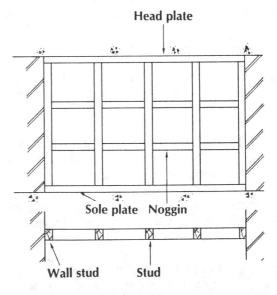

Stud partition

Subframe An outer timber-frame in which metal window or door frames can be fixed.

Subsidence A sinking of the ground, often causing foundation movement.

Subsill An extension to a timber windowsill or a second sill fitted below the main windowsill. Both are intended to throw rainwater clear of the building.

Substructure All structure below ground level, up to and including the ground floor slab and damp proof course. Its function is to receive the loads from the superstructure and transfer them safely down to a load bearing layer of earth.

Subway A pedestrian walkway below ground level to permit passage under a road, other obstruction or to connect buildings. May also be for maintenance purposes where it would contain services. Also an underground railway.

Superstructure All structure above the substructure both externally and internally. Its function is to act as the external envelope, in addition to receiving the dead and imposed loadings and transferring them down onto the substructure.

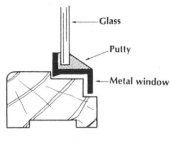

Subframe

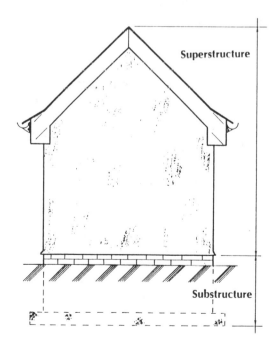

Substructure/superstructure

Surround   One material placed around another, such as the trim around an opening.

Suspended ceiling   Normally a false ceiling.

Suspended floor   A floor that spans between supports rather than a solid floor which is built against the earth. (141)

Suspended timber ground floor   A ground floor consisting of timber joists and boarding supported by sleeper walls built off the oversite concrete. Also termed a hollow floor.

Suspended timber upper floor   An upper floor consisting of a number of bridging joists supported at either end by load bearing walls. Joists are covered on their upper edges with boarding or sheeting to provide the floor surface and on their underside with plasterboard to form a ceiling. Solid or herringbone strutting is used at mid-span to stiffen the whole floor. *See also* Single, Double and Framed floors and Trimming.

Sussex garden wall bond   Flemish garden wall bond.

Swan neck   A component having an S-shaped bend.

Swept valley   A valley formed using wedge-shaped slates or tiles which gives a continuous course around the two intersecting roof surfaces and avoids the use of a valley gutter. Also termed a laced valley.

Swing door   A double action door.

Swinging jamb or post   The hanging jamb or post of a door or gate.

System building   A method of building that utilizes a limited number of large, factory prefabricated components. The intention of this is to reduce costs through mass production and also to speed on-site construction.

Tail piece   A trimmed joist.

Tanking   A waterproof membrane laid over a basement floor and up the walls. It is held in position, and protected from damage, by a loading coat over the floor and a half brick skin on the outside of the walls.

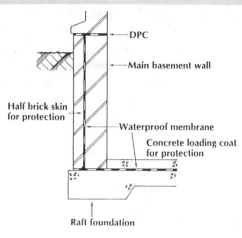

Tanking

Tee bar   A proprietary suspended or false ceiling system.

Temporary bench mark   A datum peg established on site from which all required levels can be obtained. Normally has a level value which is related to an ordnance bench mark.

Tenon   The end of a framing member that is reduced in thickness to enable it to be inserted into a mortise in another member to which it is to be joined. The thickness of a tenon should be approximately one-third the thickness of the timber to be joined and its width not more than five times its thickness. *See also* Double, Twin, Haunch, Stub and Tusk tenons.

Three-quarter bat   A three-quarter brick in length.

Three-quarter header   A three-quarter brick in width.

Threshold   The horizontal timber sill of an external door frame. (94)

Throat   A drip groove. Also the undercut groove run around the rebate of window and external doorframes, to prevent the entry of wind-assisted rainwater.

Through tenon   A tenon that passes right through the mortised member as opposed to a stub tenon that stops short.

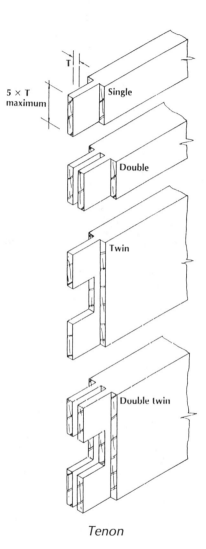

Tenon

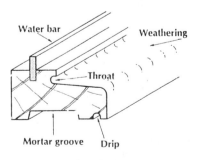

Threshold

Tie   A structural member that is in tension e.g. a ceiling tie or joist that ties the feet of rafters together. (78, 80)

Tile hanging   Tiles used as a wallcladding to provide weathering for the structure.

Tilt   To raise at an angle. Also termed cant.

Tilting fillet   A triangular section length of timber fixed to the eaves of a roof to provide a bedding for the undercloak; behind the back gutter to a chimney stack for the same purpose. Also fixed on top of a flat roof where it abuts a wall, or upstand fascia to ease the turning up of the roofing felt around the corner. (92)

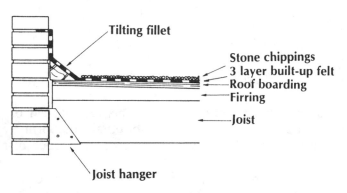

Tilting fillet

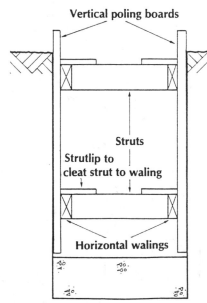

Timbering to the sides
of a foundation trench

Timbering

Timbering   Any timber work used to provide temporary support, such as formwork, shoring and centres for arches. More specifically applied to the temporary support to the sides of trench excavations whilst construction work is being carried out.

TMB   Temporary bench mark.

Toe   The lower portion of a closing stile. *See also* Heel.

Toe nailing   Skew nailing.

Tongue   A central projection on the edge of a board formed by rebates on either side. It is inserted into the groove of the adjacent board, forming a tongue and groove joint. *See also* Cross tongue and Bared faced tenon.

Tongue and groove   An edge joint between adjacent timber boards using tongues and grooves, such as tongue and groove flooring or matchboarding. *See also* Vee jointed.

Tooled joint   *See* Jointing.

Toother   The stretcher brick that projects in toothing out.

Toothing out   Stretchers left protruding at the end of a wall, like teeth on alternate courses, as a bond for future work.

Top cut   A plumb cut.

Top rail   The uppermost horizontal member of door, window and framed wall panelling.

Transom   The intermediate horizontal member of a window or doorframe between the head and sill or threshold. *See also* Mullion.

Tread   The horizontal member of a step. Can be classified according to its shape e.g. parallel tread, tapered tread. *See also* Riser and Winders. (149)

Treenail   A timber draw pin or dowel.

Trench   A groove cut across the grain of a timber member. Also a long narrow hole dug in the ground to accommodate foundations or pipes etc.

Trial hole   A hole dug or bored on site during a provisional site investigation to determine the nature of the ground.

Trim   The collective term for skirtings, architraves and cover fillets etc.

Trimmed   A bridging joist that has been cut short to form an opening in a suspended timber upper floor. The trimmmed ends are supported by a trimmer. *See also* Trimming.

Trimmer   A joist placed at right angles to the bridging joist in order to support the cut ends of the trimmed joists. Trimming joists are used to support the ends of the trimmer to form openings in suspended timber upper floors mainly for stairs. *See also* Tusk tenon.

Trimming   The framing of joists or rafters where openings occur in floors and roofs using trimmed, trimmer and trimming.

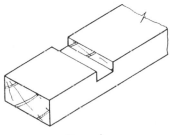

Trench

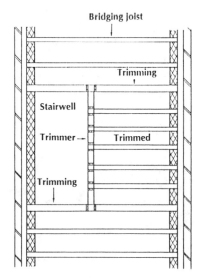

Trimming

Trimming joist   A joist spanning the same direction as the bridging joist but supporting the end of a trimmer joist. Used to form openings in suspended timber floors, mainly for stairs. *See also* Trimmed.

Tripple floor   *See* Framed floor.

Truss   A structural frame designed to act as a beam e.g. it spans an opening. *See also* Roof truss.

Trussed beam   A made-up beam in the form of a truss.

Trussed purlin   A trussed beam used as a purlin.

Trussed rafter   A pitched roof component consisting of a pair of rafters formed into a truss, of a fink or fan pattern. Joints are normally made using nail plates. *See also* Roof truss. (96)

Tuck pointing   A recess formed in horizontal mortar joints and filled with lime putty. This projects slightly from the face producing decorative white lines. (115)

Tumbling in   The intersection between sloping courses of brickwork and horizontal courses, such as at a gable wall coping.

Turning piece   A solid piece of timber cut to the shape of a flat or segmental arch. Used to provide temporary support during construction. *See also* Arch centre.

Turret roof   A pitched roof to a tower or similar slender structure.

Tusk   The projecting part of toothing.

Tusk nailing   Skew nailing.

Tusk tenon   A structural framing point used in traditional floor construction between the trimmer joist and the trimming. Mainly superseded by joist hangers. Consists of a tenon and bottom tusk with a splayed top shoulder that passes through a mortise and is keyed with a wedge driven through a hole in the tenon.

Twin tenon   Two tenons arranged within the width of a piece of timber, separated by a haunch, one following the other. Normally used for jointing the middle and bottom rails of doors to the stiles. *See also* Double tenon. (157)

Uncoursed   Random rubble walling. Walls built with varying size units so that a common bed course is impractical.

Undercloak   The lower course of slate or plain tiles bedded at the eaves or verge. At the eaves they form the double thickness, whereas at the verge they provide a slight upwards slope towards the roof to restrict rainwater run-off. *See also* heading under Services and finishes.

Underpinning   An additional support structure positioned under the existing foundations of a building. May be required due to foundation settlement, additional loading of building or the excavation of adjacent ground below the level of the existing foundation.

Upper floor   The floor levels of a building above the ground floor. Also termed suspended floor. Normally of timber in house construction and concrete for other buildings.

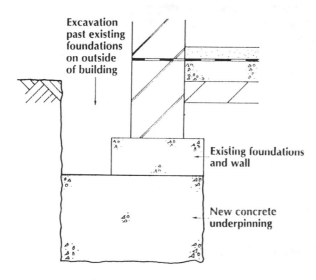

Underpinning

Valley The intersection of two pitched roof surfaces at an internal corner. *See also* Hip.

Valley gutter The normally sheet metal gutter which follows the line of the valley rafter. *See also* Swept valley.

Valley jack A cripple rafter.

Valley rafter The rafter used where two pitched roofs meet at an internal angle. It provides a fixing for the cripple rafters, supports the valley gutter and transfers loads to the wall. *See also* Hip rafter and Lay board.

Vapour barrier An impermeable barrier used on the warm side of a structure to prevent the passage of water vapour from within the building into the structure, where it could result in interstitial condensation. *See also* Moisture barrier. (164)

Vault A room, normally below ground level, for keeping valuables. Also a room or other space roofed over with arched masonry. Traditionally termed an undercroft.

Vault light A pavement light.

Vee joint Small chamfers on the meeting edges of matchboarding. Forms a V-shape on assembly. Intended to mask the effect of any subsequent shrinkage.

Vent light A small opening casement window above a transom. Also termed a fanlight or nightvent, normally top hung outward opening, but can be bottom hung inward opening when they are called hopper or hospital lights.

Ventilation brick An air brick.

Verge The termination or edge of a pitched roof at the gable. *See also* Bargeboard. (104)

Vergeboard A bargeboard.

Vertical sash A sash window that slides vertically rather than horizontally.

Vertical shore A pair of vertical struts propping up a needle. Used to provide temporary support to a wall.

Vertical tiling Tile hanging, forming an exterior cladding.

Vestibule *See* heading under Architectural style.

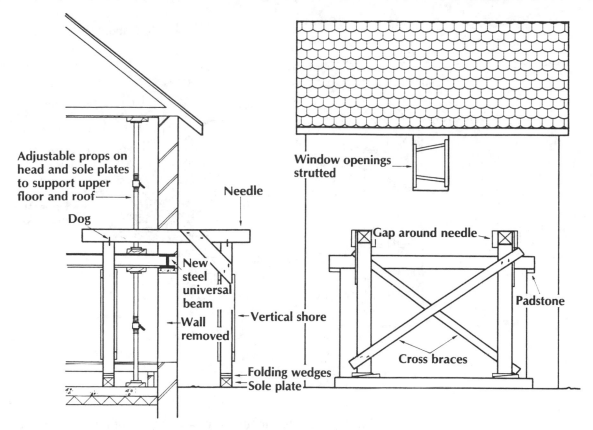

Vertical (dead) shoring

Voussoir The wedge-shaped bricks or masonry of an arch. *See also* Intrados and Extrados.

Wagtail The hanging, thin section of timber that separates the weights in box frame sash windows. Also termed parting or mid-feather. (73)

Wainscot or wainscoting Timber panelling up to dado height, traditionally made from oak.

Walking line A setting-out line for stairs with winders, where the risers and goings have to conform with the Building Regulations. This is in the centre of flights less than 1 m wide and 270 mm in from either edge for wider ones.

Wall The vertical enclosing and dividing elements of a building. May be load or non-load bearing. External walls form part of the external envelope. May be either solid, framed, or cavity.

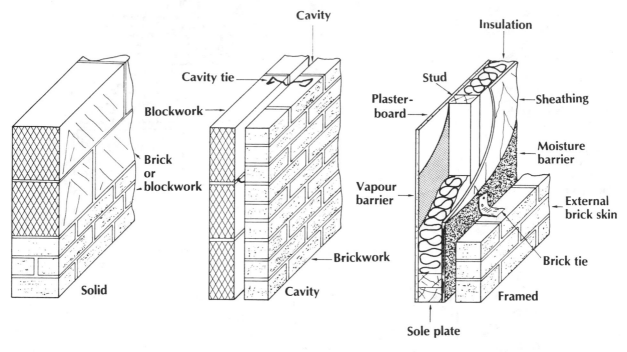

Wall

Wall hung   A component fixed or hung at high level on a wall e.g. a wall hung cupboard or a wall hung boiler.

Wall panel   A prefabricated walling section, often used as an infill between a structural framework.

Wall piece   A vertical timber fixed to the face of a wall to provide a bearing for raking and horizontal shores.

Wall plate   A horizontal piece of timber fixed to the top of walls in order to distribute the load from, and provide a bearing and fixing point for, joists and rafters. (70, 80)

Wall string   The string of a stair next to the wall. *See also* Outer string. (149)

Warm roof   A roof constructed with its insulating layer above the roof space rather than a cold roof which is at ceiling level.

Waste or Waist   The narrowest part of an object e.g. the waste of a concrete stair flight is the measurement between the sloping soffit and the internal tread and riser intersection; an allowance (often 10 per cent) made in estimating quantities of materials that will be unused; also any building rubbish. *See also* heading under Services and finishes.

Water bar A metal or plastic member incorporated in a groove either between a timber sill and concrete subsill or in the top of a timber threshold to prevent water penetration. Also a rubber or plastic member embedded into concrete movement or construction joints of water-retaining structures. (157)

Water table The natural level of ground water on a site. Also a projection near the base of a wall intended to throw rainwater clear of the building.

Weather bar A water bar.

Weather fillet A triangular section of cement mortar, applied where two building elements join at right angles. Its purpose is to prevent moisture penetration. Also termed haunching.

Weather or weathering The slope given to external horizontal surfaces to permit rainwater run off; the change of colour, texture or deterioration of a material after exposure to the weather; also the waterproofing of the external envelope. (157)

Weather strip A strip of material (metal, rubber, foam or nylon pile) normally proprietary, that is fixed around the rebate of doors and windows in order to restrict the passage of draughts and rainwater.

Weatherboard A moulding fixed to the bottom rail of exterior doors. It throws water that has run down the face of the door clear of the water bar and well onto the threshold weathering. Also one piece of weatherboarding.

Weatherboarding An external cladding of normally horizontal timber boarding; usually either tongue and groove, shiplap which is rebated, or feather edge, also termed clapboard which is simply overlapped.

Weathered joint A mortar joint finished by weathered pointing or jointing. (115)

Weathered pointing or jointing The finishing of a wall's mortar joint with a trowel so that the upper edge is set back while the bottom edge is flush with the face. This gives a weathered surface to throw water away from the joint. (115)

Wedge A tapering piece of timber or metal that is used to tighten one member against another, as in a mortise and tenon joint, or folding wedges.

Weep holes   A small drain hole for moisture. This may take the form of a pipe through retaining walls to prevent the build-up of groundwater or a hole drilled through timber windowsills to enable condensation to drain to the outside or vertical perp joints on the outer leaf, raked free of mortar above cavity trays.

Welsh arch   A small opening in brickwork spanned by a stretcher splayed at both ends and supported by two similarly splayed end, brick corbels. Also termed a jack arch.

Welt   A seam in flat metal roof coverings where the edges are folded and dressed down. Also called single and double welts depending on the number of folds.

Welted drip   The finish at the eaves or verge of a felt-covered flat roof, where the felt is fixed.

Wheel step   A winder.

Whole brick wall or one brick wall   A wall, the thickness of which is one brick (approximately 215 mm). *See also* Half brick wall. (107)

Wind or winding   See In wind.

Wind filling   Beam filling.

Winder   The tread of a stair that is triangular or tapered. Used to change the stair's direction. *See also* Kite winder.

Winding stair   A helical stair.

Window   A glazed opening in a wall used to admit daylight and air or give occupants an outside view. *See also* Casement, Sash and Pivot windows.

Window back   The panelling or wall internally below a window. Traditionally often concealed a lifting shutter.

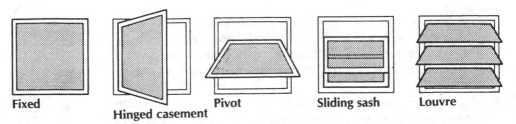

Fixed   Hinged casement   Pivot   Sliding sash   Louvre

Windows

Window bar   A glazing bar.

Window bead   A staff bead or glazing bead.

Window board   A horizontal board fixed internally at sill level, providing a shelf and finishing the top of the wall. *See also* Stallboard.

Window frame   The part of a window that is fixed into the wall opening and receives the casements or sashes.

Windowstool   A windowboard. *See also* Stallboard and Stall riser.

Wing light   A light adjacent to a door but not continuing for the full door height. *See also* Flanking window and Sidelight.

Wood block floor   A parquet floor.

Wreath   A handrail portion that is curved in both plan and elevation. Mainly used for geometric stairs.

Wreath string   The curved portion of a geometric stair string.

Y

Yoke   An arrangement of members used in formwork, which encircle beam or column forms to secure them together and prevent movement.

Yorkshire bond   Monk bond.

Yorky   A slate having a curved bed.

Z

'Z' bar   A proprietary suspended or false ceiling system.

3
DOCUMENTATION, ADMINISTRATION and CONTROL

Abstracting The process of gathering together all related information prior to writing a bill of quantities.

ACAS The Advisory, Conciliation and Arbitration Service. This acts as an independent arbiter to resolve deadlocked disputes between two parties.

Agent The building contractor's resident on-site representative and leader of the site workforce. He or she is directly responsible to the contracts manager for the day-to-day planning, management and building operations. Also known as the site manager. *See also* General foreman and Contracts manager.

Agrément Board A body which is concerned with the testing, assessment and certification of new or innovatory products/processes for the building industry. Issues Agrément Certificates.

Agrément Certificate A statement issued by the Agrément Board which gives an independent opinion of the performance of a building product, component, material or system, when used or installed in a specified manner.

Apprentice The learner of a craft who is indentured to a master for a set number of years in order to become a craftsperson.

Apprenticeship The agreement between an apprentice and his or her master, known as the deed of apprenticeship or indentures.

Approximate quantities A provisional estimate of materials and labour to be allowed for. Also known as provisional quantities.

Arbitration The process by which a dispute between two parties that is in deadlock can be resolved. This involves appointing an independent arbiter to decide between them. *See also* ACAS.

Architect A person who designs and supervises the construction of buildings and other structures. He or she is the client's agent and is considered to be leader of the building team.

Architect's instruction *See* Variation order.

Artisan An outdated term for a craft operative or tradesman.

A

Assembly drawings Working drawings that show, in detail, the junctions between the various elements and components of a building.

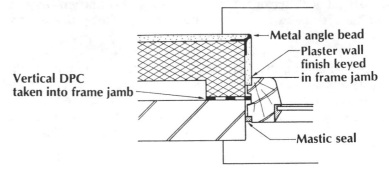

Vertical DPC taken into frame jamb

Metal angle bead

Plaster wall finish keyed in frame jamb

Mastic seal

Assembly drawing (window jamb)

B

Banker A stonemason who cuts and smooths building stone. *See also* Fixers and Freemason.

Bar chart A building programme where the individual tasks are listed in a vertical column on the left-hand side of the sheet and a horizontal time scale is included along the top. The target times of the individual tasks are shown by a horizontal bar. Plant and labour requirements are often included along the bottom of the sheet. Also known as a gantt chart.

Bill of quantities or BOQ A document prepared by the quantity surveyor. It gives a description and measure of quantities of labour, materials and other items required to carry out a building contract. It is based on the drawing specifications and schedules and forms part of the contract documents.

Billing The process of writing a bill of quantities. *See also* Quantity surveyor.

Block plan A location working drawing that shows the position of the proposed building and the general layout of roads, services and drainage etc. on the site. (175)

Bonus An extra monetary payment for working in excess of that normally required.

Brickie A slang term for a bricklayer.

Task	Week comm.	5 Sep.	12 Sep.	19 Sep.	26 Sep.	3 Oct.	10 Oct.	17 Oct.	24 Oct.	31 Oct.	7 Nov.	14 Nov.	21 Nov.	28 Nov.	5 Dec.	12 Dec.
	Week no.	1	2	3	4	5	6	7	8	9	10	11	12	13	14	15
Site preparation																
Setting out																
Excavate foundations and drains																
Concrete foundations																
Lay drains																
Brickwork to DPC																
Hardcore/concrete to ground floor																
Brickwork to first floor																
First floor joists																
Brickwork to eaves																
Roof structure																
Roof tile (SC)																
Carpentry and joinery					1st fix		2nd fix									
Plumbing (SC)						1st fix		2nd fix								
Electrical (SC)						1st fix		2nd fix								
Internal partitions																
Water, electric, gas, and telecom mains (SC)																
Plastering																
Decoration and glazing						Glazing		Decorations								
Internal finishing																
External finishing																

Contract completion 6 Dec.

Labour requirements

	1	2	3	4	5	6	7	8	9	10	11	12	13	14	15
GD	3	2	3	1	1	1	1	1	1	1	1	1	1		
DL		2													
BL			4/2L	4/2L	2	2/1L									
CJ				2	2	2		2							

Plant requirements

JCB															
Mixer												1			
Scaffold															

Bar chart

Notes

GL general operative
DL drain operative
BL bricklayer
CJ carpenter and joiner
(SC) Subcontractor
4/2L 4 craft operatives and 2 labourers
—— Target time

PSB design

Job title: PLOT 3 ROYAL LODGE
Drawing title: PROGRAMME
Job no.: 202/89
Drawing no.: 1/111
Scale | Date 25/7/89 | Drawn PSB | Check cab.

ITEM	DESCRIPTION	QUNTY	UNIT	RATE	AMOUNT
	PRELIMINARIES				
	NAME OF PARTIES				
	CLIENT: Ms T. JOYCEE ARCHITECT: PSB DESIGN				
	25 DAWNCRAFT WAY				
	WHILTON				
	NORTHANTS				

ITEM		QUNTY	UNIT	RATE	AMOUNT
	PRELIMINARIES (cont)				
	PROVISIONAL SUMS				
A	PROVISIONAL SUMS REFERRED TO IN THESE BILLS OF QUANTITIES ARE FOR WORK OR COSTS WHICH CANNOT BE ENTIRELY FORSEEN, DETAILED OR DEFINED AT THIS PRE-CONSTRUCTION STAGE. THEY ARE TO BE EXPENDED AS DIRECTED BY THE				

ITEM		QUNTY	UNIT	RATE	AMOUNT
A	BRICKWORK (cont)				
B	NO BROKEN OR CRACKED BRICKS OR BATS SHALL BE USED EXCEPT FOR BONDING PURPOSES. THE BRICKWORK SHALL RISE FOUR COURSES TO 300mm. ALL BRICKS TO BE WETTED PRIOR TO USE.				
B	ALL WALLS SHALL BE BUILT UP REGULARLY ALL				

ITEM		QUNTY	UNIT	RATE	AMOUNT
	FITTINGS				
	KITCHEN FITTINGS				
A	PROVIDE THE P.C. SUM OF TWO THOUSAND, ONE HUNDRED AND SIXTY POUNDS (£2160.00) FOR SUPPLY AND DELIVERY TO SITE OF THE FOLLOWING KITCHEN FITTINGS FROM A SUPPLIER TO BE NOMINATED BY THE ARCHITECT.				2160 00
B	ADD FOR EXPENSES AND PROFIT			%	
	FIT IN POSITION THE FOLLOWING KITCHEN FITTINGS, INCLUDING; THE ASSEMBLY OF COMPONENT PARTS AS REQUIRED; PLUGGING AND SCREWING TO BLOCKWORK WALLS; FITTING AND FIXING WORKTOPS; GENERAL EASING AND ADJUSTING AS NECESSARY.				
	FLOOR UNITS (WORKTOPS MEASURED SEPARATELY)				
C	SINK BASE UNIT 600mm x 1000mm	1	No		
D	BASE UNIT 600mm x 1000mm	2	No		
E	BASE UNIT 600mm x 1000mm	2	No		
	WALL UNITS				
F	DOUBLE UNIT 900mm x 1000mm	2	No		
G	BROOM CUPBOARD 600mm x 600mm x 2100mm	1	No		
	WORKS TOPS				
H	LAMINATE TOP 600mm x 3000mm SCREWED TO BASE UNIT (MEASURED SEPARATELY)	2	No		
J	SLATTED SHELVING TO BROOM CUPBOARD 19mm x 44mm PAR SOFTWOOD AT 60mm CENTRES SCREWED TO BEARERS.	1.5	M²		
K	INCLUDE THE PROVISIONAL SUM OF ONE HUNDRED AND FIFTY POUNDS (£150.00) FOR CONTINGENCIES.				150
	CARRIED TO COLLECTION			£	

Bill of quantities

Bricklayer The skilled craft operative who works with bricks and mortar to construct all types of walling.

Bricklayer's labourer The general operative who carries bricks and mixes and carries mortar to the bricklayers. *See also* Hod carrier.

British Standards A publication of the British Standards Institution (BSI), British Standard Specification (BSS) deal with materials and components. British Standards Codes of Practice (BSCP) cover design and workmanship.

British Standards Institution (BSI) The British organization which produces voluntary standards and encourages their use in consultation with all interested parties including manufacturers and users. The scope of their publications include: glossary of terms; definitions and symbols; methods of testing; methods of assembly or construction; specification for quality; safety; performance or dimensions. *See also* British Standards, ISO and DIN.

Brush hand A painter, a term often applied to one who is not fully skilled.

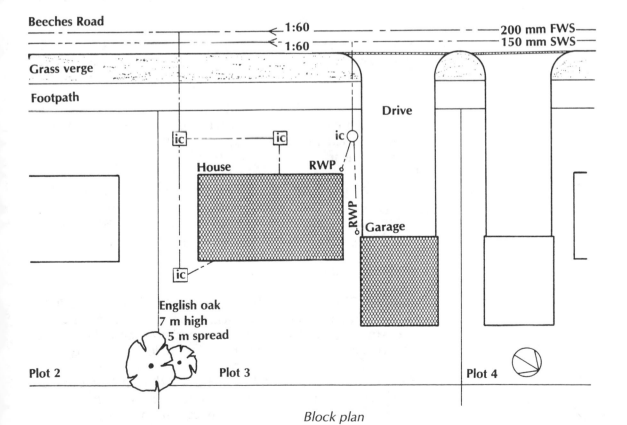

Block plan

BSCP *See* British Standards.

BSI *See* British Standards Institution.

BSS *See* British Standards.

Builder's labourer A general operative.

Building The construction, maintenance and adaptation of buildings ranging from office blocks, industrial complexes and shopping centres to schools, hospitals, recreation centres and dwellings. *See also* heading under Architectural style.

Building control The legislation under which the building industry operates. This can be divided into three main areas: planning permission; building regulations and health and safety controls.

Building craft One or all of the craft operative level occupations in the building industry.

Building industry The general term used to include building, civil engineering, mechanical engineering and electrical engineering.

Building inspector An employee of a local authority who approves and inspects building work. Also known as building control officers, and district surveyors.

Building notice A method of complying with building regulations. Involves giving a local authority limited information of intended building work forty-eight hours before commencement.

Building operative A person who carries out the actual physical building work. Known as general and specialist building operatives according to the type of work carried out. *See also* Labourer and Craft operative.

Building owner *See* Client.

Building Regulations Building control legislation that states how a building should be constructed, altered or extended to ensure safe and healthy accommodation and the conservation of energy. Administered by the local authority.

Building Research Establishment (BRE) A government-backed organization that carries out research into practical building problems. Gives advice on the preparation of

British Standards and building regulations and answers technical enquiries.

Building surveyor A person who determines positions for buildings, roads and bridges etc. or who is concerned with the administration of new building works, adaptation and maintenance. *See also* Quantity surveyor.

Building team The team of professionals who work together to produce the required building or structure. Consists of the following parties: client, architect, quantity surveyor, specialist engineers, clerk of works, local authority, health and safety inspector, building contractor, subcontractor and suppliers.

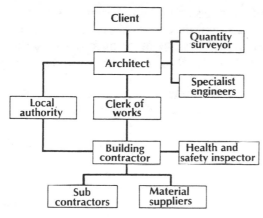

Building team

Building trade One, or all, of the operative-level occupations of the building industry. *See also* Craft operative and Building operative.

Building work The work carried out by the building industry.

Butcher A site slang term for a carpenter, often used for one of limited skill. *See also* Cowboy.

C

Cabinet maker A joiner who specializes in furniture and cabinet construction.

Carpenter The skilled craft operative who works with timber and other allied materials. Mainly works on site, as opposed to a joiner who mainly works at the bench.

Certificate of practical completion *See* Practical completion.

Chancer A person who undertakes the work of a craft or specialist operative without having undertaken a training period or apprenticeship i.e. the person takes a chance on being able to do the work.

Chargehand An assistant to the craft foreman who supervises a subsection of the work or a small team of craft operatives in addition to carrying out the skilled physical work of their craft. Also known as a working foreman.

Chippy A site slang term for a carpenter. *See also* Butcher and Cowboy.

Chop Getting the chop. Having your employment terminated. Also known as the sack.

CI/SfB classification An international classification used for indexing building information. It was originated in Sweden by their building committee and termed SfB (Samarbetskommitten for Byggnadsfragor). The version used in the UK is prefixed CI meaning Construction Industry. The CI/SfB system is based on five tables from 0 to 4 covering the physical environment, building elements, constructions, materials, activities and requirements. It is standard practice to include the standard reference code box on the top right hand corner of all information that is to be filed. This includes bill of quantities, specifications, schedules, working drawings and manufacturers' technical brochures etc.

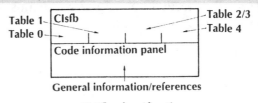

CI/sfb classification

Clerk of works (COW) The architect's/client's on-site representative responsible for checking that the contractor carries out the work in accordance with the drawings and other contract documents.

Client   The actual person (or organization) who has a need for building work. The client is responsible for the overall financing of the work and either directly or indirectly employs the entire building team.

Code of Practice   A British Standard Code of Practice (BSCP). A publication of the British Standard Institution that covers the design and workmanship of whole building processes.

Competitive tendering   Tendering for a building contract in competition with other contractors. May be open or selective tendering. *See also* Negotiated contract and Speculative building.

Component drawings   Working drawings that either show a standard range of components e.g. doors, windows, sanitaryware etc. or detail drawings that show all the information required to manufacture a particular component.

Conditions of contract   The standard form of contract which both parties sign, such as the Joint Contractors Tribunal (JCT) or the Building Employers Confederation (BEC) Forms of Contract. The actual form of contract used will

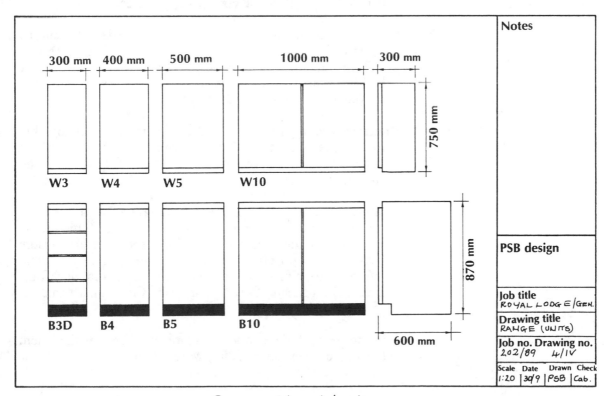

Component (range) drawing

depend on the type of client, the size and type of work and the extent of the contract documents but each will include the rights and obligations of all parties and details of procedures for variations, interim payments, retention and defects liability period.

Confirmation notice   Written confirmation of verbal instructions given by the architect or clerk of works for daywork or variations.

Construction   The process of building.

Construction process or stage   The actual physical tasks and administrative processes of building.

Construction Regulations   Safety regulations made under the Factories Act which are specific to construction operations.

Consultant   Specialist engineers who are consulted on building design aspects e.g. civil, structural and service engineers.

Contingency sum   A provisional sum of money to be included in a tender for work which has not yet been finally detailed or to cover the cost of any unforeseen work.

Contract   Legal agreement between two parties. A building contract is an agreement between the building contractor and the client. The contractor agrees to carry out the building work and the client agrees to pay a sum of money for the work. *See also* Indentures.

Contract documents   The various documents that together form the legal contract. These normally consist of working drawings; specification; schedules; bill of quantities and conditions of contract.

Contract planning   A process undertaken by a building contractor, it involves producing a programme. This should show the sequence of work activities, the interrelationship between the different tasks and when, and for what duration, resources such as materials, plant and workforce are required on site. *See also* Bar chart, Critical path analysis and Networks.

Contractor   A builder who enters into a contract with a client to carry out certain building works in accordance with the contract documents.

Contracts manager   The supervisor of the site management teams on a number of building contracts. The contracts manager has overall responsibility for planning management and building operations. He or she will liaise between the site agent and head office staff.

Costing The process of estimating or working out the cost of the labour, plant and materials for carrying out an item or unit of work. This cost will be the unit cost that is entered in the bill of quantities.

COW *See* Clerk of works.

Cowboy A site slang term for a craft operative of limited skill, a chancer or one who works on the lump.

Craft A skilled occupation. *See also* Craft operative.

Craft foreman The person who works under the general foreman to organize and supervise the work of a specific craft e.g. foreman bricklayer and foreman carpenter etc. *See also* Ganger and Chargehand.

Craft operative A skilled person who carries out the actual physical building work e.g. bricklayer, carpenter, electrician, joiner, painter, plasterer and plumber.

Critical path analysis A form of network where the estimated time required to carry out each building activity is shown. The duration of those activities which are critical to the completion of the project are highlighted, thus giving the critical path. Any alteration in the duration of these critical activities will affect the overall project time. Also shown on a critical path network are activities having a float time. The duration of these can be extended without affecting the overall project time.

Daywork Work which is carried out by a contractor without an estimate. Normally work extra to the main contract for which a set hourly or daily rate is payable.

Decorator A skilled craft operative who can paint and hang wall coverings. *See also* Painter and Paper hanger.

Deed A legal contract such as an indenture or deed of apprenticeship.

Defects liability period A set period of time, often six months after practical completion, to allow any defects to become apparent. After rectifying any defects to the architect's satisfaction the contractor will become entitled to payment of the retention.

Demolisher A person who carries out demolition work.

Demolition The process of pulling down a structure or building.

Detail drawings *See* Component drawings.

Dimension paper Paper used by the quantity surveyor for taking off.

Dimensional coordination The standardization of a range of dimensions for building components, assemblies and buildings. This uses the basic module of 100 mm or its multiple for both vertical and horizontal dimensions. Also termed modular coordination.

DIN The German equivalent of British Standards; stands for Deutsche Industrie Norm or German Industry Standard.

Direct labour The employment of craft and general operatives by the client, leaving out the building contractor. Many local authorities have their own direct labour departments to carry out building work. *See also* Lump and Subbies.

Dirty money An extra payment to an operative for carrying out work of an unpleasant nature.

District surveyor or DS A building inspector.

Drummer The person who makes the tea on site, often the youngest apprentice or trainee.

DS District surveyor.

Dyker A stonemason who builds dry stone boundary walls.

Easement A right in law to use or have access to land which is not one's own such as to run drainage pipes under it or erect a scaffold on it to effect repairs to one's own property.

Electrician The skilled operative who installs and maintains electrical systems.

Engineer *See* Site engineer.

Estimating The process of working out the cost of carrying out a certain item of work. *See also* Estimator, Costing and Bill of quantities.

Estimator The person who carries out estimating work.

Extras Additional work not included in the contract. Must be ordered or agreed by architect in writing, normally by the issue of a variation order.

Factory inspector A health and safety inspector who represents the factory inspectorate. His or her purpose is to ensure that government legislation concerning health and safety is fully implemented by the building contractor. *See also* Health and Safety at Work Act, Construction Regulations, Improvement notice and Prohibition notice.

Factory inspectorate A division of the health and safety executive.

Fibrous plasterer A plasterer who specializes in making and fixing fibrous plaster mouldings and decorations.

Finisher An operative who smooths or makes good a surface to provide the required standard of finish. Mostly associated with concrete surfaces.

Fitter The person responsible for the maintenance of plant items on site: crane, dumper truck and mixer etc. *See also* Pipe fitter.

Fixer A stonemason who erects stone prepared by a banker. *See also* Freemason. Can also be applied to any person who fixes a building element/component e.g. a ceiling fixer.

Foreman A person who organizes and supervises the work of others. *See also* General foreman, Craft foreman, Ganger and Chargehand.

Formworker A carpenter who specializes in the erection of formwork for casting concrete items and structure. Also known as a shuttering carpenter.

Freemason Traditionally a term applied to stonemasons who were able to work or carve freestone. *See also* Banker and fixer.

G

Ganger A foreman who works under the general foreman and organizes and supervises the work of general operatives. *See also* Craft foreman and Chargehand.

Gantt chart A bar chart.

Gasfitter The person who fits and maintains gas pipes and appliances. *See also* Pipe strangler.

General foreman The person who works under the site agent and is responsible for coordinating the work of the craft foreman, ganger and subcontractors.

General location plans A location working drawing that shows the position occupied by the various areas within a

General location plan

building and identifies the location of the principal elements and components.

General operative A person who carries out the actual physical building work which is of a general nature. Often termed labourer e.g. those who mix and lay concrete, lay drains, off-load material and assist the work of craft operatives. *See also* Specialist operative and Ganger.

General register A booklet to be completed for building operations and civil engineering. It is used to record details of site or workshop, nature of work being undertaken, any cases of disease or poisoning and the employment or transfer of young persons.

Glazier A specialist operative who cuts and installs glass.

Grainer A painter who specializes in graining e.g. the painting of a surface to look like the grain of timber or the veining of marble.

Health and Safety at Work Act The main statutory legislation covering the health and safety of all persons at their place of work and protecting other people from risks occurring through work activities.

Health and Safety Executive The body that administers the Health and Safety at Work Act.

Health and safety inspector A factory inspector.

Heating and ventilation engineer A plumber who specializes in the installation and maintenence of heating or ventilation systems. Also applied to one who designs such systems. The same as a mechanical engineer.

Hod carrier A bricklayer's labourer. General operative who carries bricks and mortar in a hod.

Improvement notice A notice issued by a factory Inspector that requires the responsible person to put right, within a specified period of time, any minor hazard or infringement of safety legislation. *See also* Health and Safety at Work Act and prohibition notice.

Improver A part-qualified operative, often a mature trainee, who works alongside a fully-skilled craft operative in order to gain further experience or to improve. Also called a mate.

Incentive A bonus or other inducement to encourage productivity.

Indentures A contract of apprenticeship. *See also* **Deed**.

Inspection certificate A certificate issued by a building inspector stating that the work is in accordance with the building regulations. *See also* Inspection notice and Interim certificate.

Inspection notice A notice given to the local authority by the builder to notify them that certain building stages have been reached so that the building inspector may visit to see if the work is in accordance with the building regulations. An inspection certificate may be issued for satisfactory work.

```
Beeches Borough Council

Building Inspection Notice


Plan no_____        Date————————
Nature of work_____
Address of work_____
The undermentioned work will be_____
       ready for inspection on
_____Signature of builder
_____Address of builder
  (1) Commencement         (5) Oversite concrete

  (2) Foundation excavations (6) Drains test

  (3) Concrete foundations   (7) Backfilling drains

  (4) DPC                    (8) Completion

  Note: Strike out those not applicable
```

Inspection notice

Inspector *See* **Building inspector** and **Factory inspector**.

Interim certificate A certificate issued by the architect which authorizes the client to make an interim payment to the contractor based on the interim valuation.

Interim payment A monthly, or other periodic payment, made to a contractor by the client based on the quantity surveyor's interim valuation of the work done and materials purchased by the contractor.

Interim valuation The value assessed by the quantity surveyor of work done and materials purchased by the contractor since the previous interim payment. *See also* Interim certificate.

Jerry builder Traditionally applied to building speculators who built houses between the First World War and the Second World War using shoddy materials and short-cut methods to secure a quick profit.

Jobbing builder A builder who undertakes small works mainly of a repair or maintenance nature. Often the craft operatives who work for such a builder will be expected to undertake work of another craft e.g. a carpenter might be required to refix the guttering after repairing a fascia board.

Joiner The skilled craft operative who makes joinery at a bench or who installs it on site. *See also* Carpenter.

Labour The actual physical building work. *See also* Labour only and Operative.

Labour only A method of contracting to build where the builder supplies the labourforce and the client supplies the building materials.

Labourer *See* General operative.

Levy A tax applied to building contractors according to the size of their workforce in order to provide a fund to finance training and education of building employees.

Local authority The district or county council who are responsible for local affairs. (Some districts have the ceremonial title of borough or city.) Scotland usually has regions/ districts. Amongst other matters they have the responsibility of ensuring that building works conform to planning requirements and building regulations. *See also* Building inspector.

Location drawings Scale working drawings that show the plans, elevations, sections, details and locality of proposed building. *See also* Block plans, Site plans and General location plans.

Lump Working on the lump. A slang term applied to a person who receives money from a contractor for work done without any deduction for tax and insurance etc. hence the person receives the whole lump.

Main contractor   The building contractor who enters into a contract with the client as opposed to a subcontractor who enters into a contract with the main contractor to undertake a specialist part of the work.

Maintenance period   The defects liability period.

Mason   A stonemason.

Mate   A general operative or part-qualified operative who assists a skilled craft operative e.g. a plumber's mate. Also known as an improver.

Measured quantities   A description and measurement of items or units of work appearing in the bill of quantities. The measurement is given in metres run, metres square, metres cube, kilogrammes etc. or just enumerated as appropriate. *See also* Measurement and Standard method of measurement.

Measurement   The process undertaken by the quantity surveyor in estimating from the drawings the amount of work to be included in the bill of quantities. Also the measuring on site of an interim valuation. *See also* Standard method of measurement.

Mechanical engineer   A plumber who specializes in the installation and maintenance of heating or ventilation systems. Also applied to one who designs such systems.

Modular coordination   *See* Dimensional coordination.

Module   A unit of size which is used either singularly or in multiples for dimensional coordination. The international basic module is 100 mm. This is used as the base unit of size for vertical and horizontal dimensions. The basic module is represented by the capital letter M. Thus 1 M = 100 mm, a 3 M wide component will be 300 mm in width. *See also* heading under Architectural style.

Navvy A general operative or labourer, especially one who digs trenches. Originally the labourers who dug the canal system – a navigation, hence navvy.

Negotiated contract Where an architect approaches a building contractor to carry out a specific project. The quantity surveyor will negotiate with the contractor to reach a mutually agreed price. *See also* Competitive tendering.

Networks A graphical method of presenting the activities or events and their interrelationship which have to be carried out in order to complete a task or project. Networks basically consist of numbered circles which are the activities in a project shown in the logical order in which they are to be carried out and arrows which indicate the direction of work flow. *See also* Bar chart and Critical path analysis. (214)

Nominated A material supplier or subcontractor named in the contract documents. This is the firm or person that must be used to supply certain stated materials or carry out a specific part of the contract.

Open tendering A form of competitive tendering where architects advertise for building contractors to submit tenders for a particular project. *See also* Selective tendering.

Operative A person who carries out the actual building work. *See also* General operative, Craft operative and Specialist operative.

Overheads A building contractor's costs which must be charged to a contract e.g. site and head office costs and the salaries of management and administration personnel.

Overtime Work carried out over and above a normal day or week's work for which an increased rate of pay is given.

Painter A skilled craft operative who paints new and existing works.

Paperhanger A painter or decorator who specializes in hanging wall coverings.

PC sum *See* Prime cost.

Piecework   Price work.

Pipe fitter   A craft operative who fits pipework e.g. a plumber, gas fitter etc.

Pipe strangler   A site slang term applied to anyone who fits pipework e.g. plumber or gas fitter etc.

Plasterer   A skilled craft operative who works with plaster and cement to finish walls, ceilings and floors. *See also* Fibrous plasterer.

Plumber   The skilled craft operative who installs and maintains pipework and appliances; also cuts and fixes sheet metal for roof coverings and flashings. *See also* Mechanical engineer.

Practical completion   The time when the building work has been completed to the client/architect's satisfaction. A certificate of practical completion will be issued and the contractor is then entitled to the remainder of the contract sum less retention.

Preamble   The introductory clauses to each section of a bill of quantities covering descriptions of materials and standards of workmanship.

Preconstruction process or stage   The process of tendering and contract planning. *See also* Construction process.

Preliminaries   The introduction to a bill of quantities or specification that deals with general particulars such as the names of parties to the contract, details of the work, a site description and any restrictions or conditions.

Price work   Work carried out by an operative for a fixed price rather than an hourly or daily rate; also termed piecework.

Prime cost or PC sum   The amount of money stated in the bill of quantities that is to be included in the tender price for work, services or materials provided by the subcontractor, supplier or local authority etc.

Progress certificate   An interim certificate.

Progress chart   A chart showing the programme of work. The actual progress of work is filled in at regular intervals and can be compared with the target progress. *See also* Bar chart.

Progress report   A report given by the general foreman/craft foreman at the site meeting on work done or reasons for any hold-ups during the period.

Project manager   *See* Agent.

Project network   *See* Critical path analysis.

Provisional sum   A contingency sum.

QS   See Quantity surveyor.

Quantity surveyor   The client's building economic consultant or accountant who prepares the bill of quantities, interim valuations and final account.

Q

Range drawings   *See* Component drawings.

Register   *See* General register.

Retention   A sum of money, often 10 per cent, which is retained by the client from cach interim valuation until the end of the defects liability period.

R

Safety officer   The person who is responsible to senior management for all aspects of health and safety. His or her role is to advise on health and safety matters; carry out safety inspections; keep records; investigate accidents and arrange staff safety training.

Saw doctor   A person who sharpens and maintains saws.

Sawyer   A person who operates a power saw.

Scaffolder   The specialist operative who erects and dismantles scaffolds.

Schedule   A contract document that is used to record repetitive design information about a range of similar components e.g. doors, ironmongery, finishes, sanitaryware etc.

S

Selective tendering   A form of competitive tendering where architects select a number of suitable building contractors with the expertise to carry out a particular project and ask them to submit tenders. *See also* Open tendering.

Sequence of operations   The order in which a building is constructed. *See also* Bar chart.

Setter out   The craft operative in a joiner's shop who sets out rods and marks up work for other operatives to shape and fit.

Shoddy   Materials or standard of workmanship that is of a low quality.

Shopfitter   A person, often carpentry based, who specializes in the fitting out of shops.

Shuttering hand   See Formworker.

Site agent   See Agent.

Site diary   A daily record of site activities, telephone calls, visitors, deliveries, delays and weather conditions, by an agent and / or general foreman.

Site manager   *See* Agent.

Skilled operative   *See* Craft operative.

SMM   Standard method of measurement.

Specification   A contract document that supplements the working drawings. It contains a precise description of all essential information and job requirements that will affect the tender price but cannot be shown on the drawings. This includes any restrictions, availability of services, description of workmanship and materials and any other requirements such as site clearance, making good and who passes the work etc.

Speculative building   Where a building company, either on their own or as part of a consortium, designs and builds a project for later sale. The builder company is speculating, or taking a chance, that they will later find a buyer. The majority of private housing is built for sale in this way.

Spread   A site slang term for a plasterer.

Standard method of measurement or SMM   A document used by quantity surveyors when writing a bill of quantities

PSB DESIGN

SPECIFICATION OF THE WORKS TO BE CARRIED OUT AND THE
MATERIALS TO BE USED IN THE ERECTION AND COMPLETION
OF A DETACHED HOUSE AND GARAGE ON PLOT 18, WHILTON
MARINE ROAD, WHILTON, NORTHANTS FOR MRS T. JOYCEE,
TO THE SATISFACTION OF THE ARCHITECT.

1.00   GENERAL CONDITIONS

1.01

1.02

1.03

1.04
1.05

2.00

2.01
2.02

2.03
2.04

17.00   GLAZING

17.01   ALL GLASS IS TO BE IN ACCORDANCE
        WITH B.S. 952.

17.02   ALL GLASS IS TO BE FREE FROM SCRATCHES,
        BUBBLES OR OTHER SIMILIAR BLEMISHES.

17.03

17.04

17.05

17.06

17.07

17.08

17.09

18.00   PAINTING AND DECORATING

18.01   ALL PAINTING WORKS TO CONFORM TO BSCP 231
        UNLESS OTHERWISE SPECIFIED

18.02

18.03

18.04

18.05

18.06

18.07

18.08

JOINERY (Continued)

19.14   PROVIDE AND FIX A DOGLEG STAIRCASE IN
        TWO UNEQUAL FLIGHTS.

        FLIGHT 1: HAVING; A RISE OF 1125MM; A
        GOING OF 1150MM; AN OVERALL WIDTH OF
        850MM; 5 No. 19MM X 296MM SOFTWOOD
        TREADS; 6 No. 9MM X 187.5MM PLYWOOD
        RISERS; TREADS AND RISERS TO BE GROOVED,
        REBATED, SCREWED AND GLUE BLOCKED TO-
        GETHER ; TREADS AND RISERS TO BE HOUSED,
        GLUED AND WEDGED TO STRINGS; 38MM X 275MM
        WALL STRING PLUGGED AND SCREWED TO BLOCK-
        WORK WALL; 38MM X 275MM OUTER STRING,
        MORTICED, TENONED AND DRAWPINNED TO 95MM
        X 95MM NEWEL POSTS AT EITHER END; NEWELS
        TO BE NOTCHED OVER LANDING TRIMMER OR
        GROUND FLOOR JOISTS AND BOLTED TO SAME
        USING 12MM DIA. X 250MM MILD STEEL COACH
        BOLTS, NUTS AND WASHERS.

        FLIGHT 2: SHALL BE AS DESCRIBED FOR FLIGHT
        ONE EXCEPT; HAVING; A RISE OF 1500MM ; A
        GOING OF 1560MM.

19.15   AFTER INSTALLATION ALL EXPOSED TREAD, RISER
        AND NEWEL SURFACES MUST BE EFFECTIVELY PRO-
        TECTED FROM DAMAGE UNTIL PRACTICAL COMPLETION/
        OCCUPATION.

Specification

so that each item is described and measured in a consistent manner.

Statutory undertaking   An organization which has a statutory duty to provide the public with certain essential requirements e.g. water/drainage, gas and electricity.

Steeplejack   An operative who builds and maintains steeples and other tall structures.

Stonemason   The skilled craft operative who works with stone. *See also* Banker, Fixer, Dyker and Freemason.

Structural engineer   A specialist consultant who will prepare drawings and calculations to assist the architect with the structural aspects of design.

Subcontract   A contract to undertake part of a main contract. Normally work of a specialist nature. *See also* Subcontractor.

Subcontractor   A specialist builder who undertakes part of a main building contract from the contractor. *See also* Nominated.

Subletting   The process of subcontracting the majority of a main contract to subcontractors.

Supplier   The person or organization supplying building materials, equipment and plant to a building contractor. *See also* Nominated.

Surveyor   *See* Building surveyor and Quantity surveyor.

Taking off   The process of compiling lists of material from drawings, rods and schedules.

Tallyman   Traditionally the person who gave out the numbered discs or tallys when people reported for work on site. These are given back at the end of a shift enabling the workforce to be accurately identified at any time.

Tendering   The production and submission of a tender price for carrying out certain stated building works based on a study of the contract documents.

Town planning   Building control legislation under the Town and Country Planning Act that exists to control the use and development of land in order to obtain the greatest possible environmental advantage, with the least inconvenience for both the individual and society as a whole.

Trade *See* Craft.

Tupper A name sometimes applied to a bricklayer's hod carrier.

Unit costing The total cost of carrying out an item of work including labour and materials e.g. the unit cost of constructing one metre of brick wall.

Variation order A written instruction from the architect to the building contractor authorizing either additional daywork or a change in the specification or drawings etc. Also known as an architect's instruction.

Waster A site slang term for a person who does little work.

Woodworking machinist A skilled person who operates any woodworking machine.

Working drawings The drawings that form part of the contract documents. These consist of location drawings, component drawings and assembly drawings.

Working foreman A chargehand.

4
GENERAL

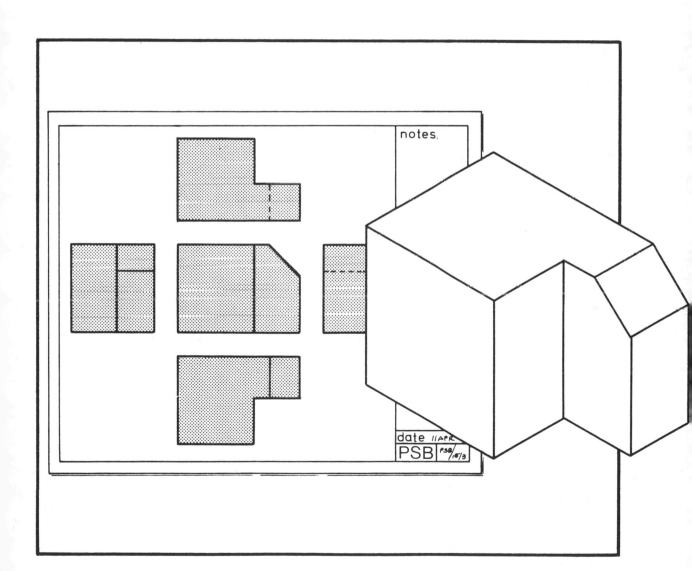

Abrasive tool A file, rasp, glasspaper or any other tool used to abrase or smooth a material.

Accommodation Buildings constructed for different purposes, such as living, working, recreation, religious activities, storage and transport.

Acute angle An angle less than ninety degrees.

Acute-angled triangle A triangle in which all three angles measure less than ninety degrees.

Addition One of the four basic mathematical operations. The sum of two numbers is determined by addition e.g. $3 + 4 = 7$. *See also* Subtraction, Multiplication and Division.

Adze A carpenter's tool traditionally used when preparing large sectioned timber, shaped like a long handled axe but has its cutting edge sideways. Used by standing astride the timber and swinging the adze between the legs. *See also* Drawknife.

Airbrush A small paint spray gun used mainly for artistic work.

Airgun A paint spray gun.

Algebra A form of mathematics that uses letters to represent numbers and signs to represent their relationship.

Altitude The perpendicular measurement of a triangle from its vertex to its base.

Angle An angle is formed when two inclined lines touch. *See also* Right angle, Obtuse angle, Acute angle, Supplementary angles and Complementary angles.

Annulus The area of a plane figure that is bounded by two concentric circles. (201)

Antilogarithm The inverse of a logarithm function e.g. to a base ten.

log 100 = 2   antilog 2 = 100

Apartment *See* heading under Architectural style.

Apex The highest point of a plane or solid figure relative to its base.

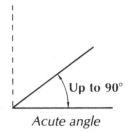

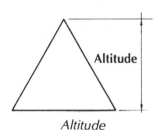

Acute angle

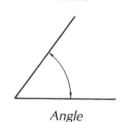

Altitude

Angle

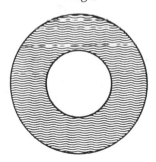

Annulus

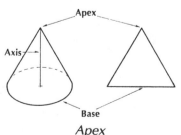

Apex

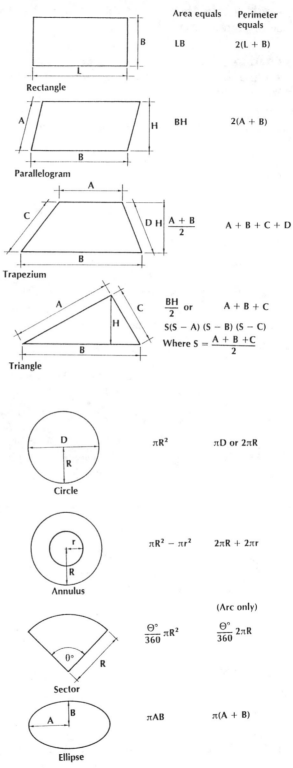

	Area equals	Perimeter equals
Rectangle	LB	2(L + B)
Parallelogram	BH	2(A + B)
Trapezium	$\dfrac{A + B}{2}$	A + B + C + D
Triangle	$\dfrac{BH}{2}$ or S(S − A) (S − B) (S − C) Where $S = \dfrac{A + B + C}{2}$	A + B + C
Circle	πR^2	πD or 2πR
Annulus	$\pi R^2 - \pi r^2$	2πR + 2πr
Sector	$\dfrac{\Theta°}{360} \pi R^2$	(Arc only) $\dfrac{\Theta°}{360} 2\pi R$
Ellipse	πAB	π(A + B)

Note: for π see Pi

Area

Approximation An inexact result which may be accurate enough for some purposes. Used in mathematics as a rough check of the expected size of an answer or positioning of a decimal point e.g. $4{\cdot}65 \times 2{\cdot}05 - 3{\cdot}85$ could be approximated to $5 \times 2 - 4 = 6$. The exact answer would be $5{\cdot}6825$.

Arc Any section of a circle's circumference.

Area The extent of a space or surface. The area of simple figures can be easily calculated using standard formula. *See also* heading under Architectural style.

Ark A chest.

Aspect The direction that a building faces. Also termed orientation.

Average A single number that represents or typifies a collection of numbers. *See also* Mode, Mean and Median.

Axis An imaginary line about which a body rotates. A plane figure will generate a solid by one revolution about its axis. Also any straight line from end to end of a plane figure or solid. (199)

Axonometric A pictorial method of drawing an object where all vertical lines are drawn vertical, while horizontal lines are drawn at forty-five degrees to the horizontal giving a true plan shape. Often used in kitchen design. *See also* Isometric and Oblique.

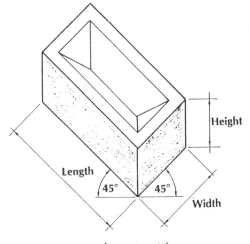

Axonometric

B

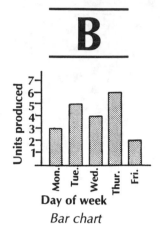

Bar chart

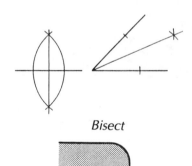

Bisect

Bottled

Ballistic tool A fixing tool that uses a ballistic cartridge, commonly termed a cartridge operated tool or nail gun.

Bar chart A graph which uses parallel bars to illustrate statistical information, the length of each bar being proportional to the quantity represented. *See also* heading under Documentation, administration and control.

Base The method of grouping used in a number system. The decimal system is based on ten. The number 45 means 5 + (4 × 10). Also the side or area on which a plane figure or solid sits. *See also* Binary.

Bat bolt A rag bolt.

Bending spring A flexible spring inserted into bendable copper tube to prevent its walls collapsing while forming bends by hand.

Binary A number system based on grouping in twos. This system only uses two digits '0' and '1'. Mainly used in computing.

Bisect To cut in half.

Bitch A steel fixing device similar to a dog but with points at right angles to each other.

Bottled A rounded edge.

Bottoming The levelling up of a foundation trench or base. Also the laying of hardcore or concrete blinding to a foundation.

Bound construction Framed construction or bonded when applied to brick and stonework.

Bowlers Large pebbles used for paving.

Brackets A mathematical symbol used to indicate in which order operations are to be carried out e.g. 7 + (50 × 2) = 107. The operation in the bracket must be carried out first to achieve the correct answer.

Brandering *See* heading under Architectural style.

Branders Battens. *See also* Brandering under Architectural style.

Breastsummer *See* heading under Architectural style.

Bridle   A trimmer joist.

Bridling   A trimming joist.

Building   The process of construction, maintenance and adaptation of buildings ranging from office blocks, industrial complexes and shopping centres to schools, hospitals, recreation centres and dwellings. Also a structure that includes an external envelope.

Built environment   The different types of construction that are required to fulfil the needs and expectations of society such as: accommodation and facilities for living, working, recreation, religious activities, storage and transport. Individually these are termed elements of the built environment.

Bungalow   *See* heading under Architectural style.

Button   A small section of wood or metal fixed through its centre with a screw, used for fastening doors and drawers etc. Also used for securing table and cabinet tops to their underframe or carcase.

Calculate   The carrying out of a mathematical process.

Calculus   The branch of mathematics concerned with the study of the behaviour of functions.

Chalet   A small, light, wooden-built dwelling. Bungalows with rooms built into their wooden roof framing are known as chalet bungalows.

Chase   An indent in a floor, wall or ceiling etc. Normally to receive a pipe cable or other building component e.g. a chase mortise.

Chord   A straight line which touches the circumference of a circle at two points but does not pass through the centre.

Circle   A plane figure bounded by a continuous curved line, every point of which is an equal distance from its centre. *See also* Concentric circles and Eccentric circles. (201)

Circumference   The curved outer line or perimeter of a circle.

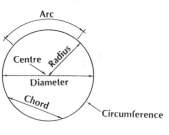

Circle

Circus *See* heading under Architectural style.

City A large or important town which is created a city by the granting of a charter, such as Birmingham. Most cathedral towns are cities such as Coventry. The business centre or original area of a large town is also called the city such as the City of London.

Civil engineering The process of construction and maintenance of public works such as roads, railways, bridges, airports, docks and sewers etc.

Civil engineering

Complementary angle Two adjacent angles that make up ninety degrees. One is said to be the complement of the other. *See also* Supplementary angles.

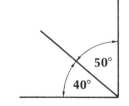

Complementary angles

Concave A hollow curved surface. *See also* Convex.

Concentric circles Circles which share the same centre but have differing radii.

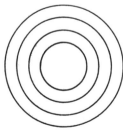

Concentric circles

Cone A solid figure described by the revolution of a right-angled triangle about one of its sides which remains fixed and is called the axis. The base of a cone is circular in shape.

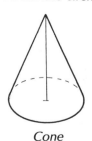

Cone

Conservation Environmental conservation is concerned with the maintenance of the environment in its entirety. This involves maintaining the cherished aspects of our heritage, such as buildings, trees, open spaces, landmarks, and the atmosphere. *See also* Pollution.

Construction The process of building.

Conurbation A merging of a group of towns and villages into one large, built-up urban area. *See also* Suburban.

Convex   A raised curved surface. The opposite of concave.

Cosine   A trigonometric function. *See* Trigonometry.

Cottage   *See* heading under Architectural style.

Crescent   *See* heading under Architectural style.

Cross-section   The plan shape achieved when a solid is cut through at right angles to its axis.

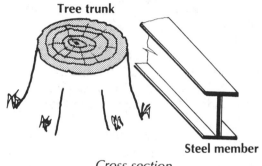

Tree trunk

Steel member

Cross section

Cube   A prism formed by six faces, all of which arc squares.

Cuboid   A prism with six rectangular faces. The opposite faces being the same size.

Cylinder   A solid figure described by the revolution of a rectangle about one of its sides which remains fixed and is called the axis. The ends of a cylinder are circular in shape.

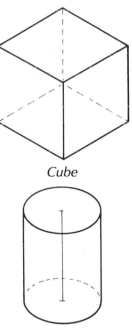

Cube

Cylinder

D

Darby A plasterer's long, two-handed, wooden float or straight edge. Used for levelling surfaces.

Data Collected information.

Decagon A polygon having ten sides.

Decimal A number system representing whole and fractional numbers in the base of 10. For example $7.5 = 7\frac{1}{2}$.

Degree A unit for the measurement of angles in which one complete rotation or circle is divided into 360 degrees or 360°. Each degree may be subdivided into sixty minutes. *See also* Temperature.

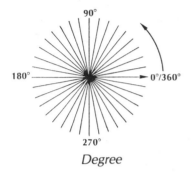

Degree

Density The system of measuring the amount of housing (known as accommodation density) or the number of people (known as population density) in a specific area of land. *See also* heading under Materials and scientific principles.

Derelict Land or buildings that have been damaged by serious neglect or other processes, which in their existing state are unsightly and incapable of use without treatment. *See also* Obsolescence and Renovation.

Detached A house or bungalow that is separate and not connected to its adjacent one. *See also* Semidetached and Terraced.

Detached

Development A drawing showing the shape of one or more faces of a solid or a truncated solid laid out flat in one plane.

Devilling float A plasterer's wooden float having nail points slightly protruding from each corner. Used for devilling i.e. the scratching of a surface to provide a mechanical key.

Diagonal A line in a polygon which joins any two non-adjacent corners or vertices.

Diagonal

Diameter A straight line (or its measurement) which passes through the centre of a circle and is terminated at both ends by the circumference.

Division One of the four basic mathematical operations, involving the sharing of one quantity into a number of equal parts e.g. $10 \div 2 = 5$. *See also* Addition, Multiplication and Subtraction.

Dog A U-shaped metal fixing device used to secure large section structural timbers together, such as a needle to a vertical shore. *See also* Bitch.

Dooking Plugging a wall to provide a fixing. The wooden plug is called a dook and the plugging chisel a dooking iron.

Drag A toothed steel plate used for levelling plaster surfaces and providing a key for the mechanical bonding of the next coat. Also a tool used to smooth ashlar stonework.

Draw knife The carpenter's tool traditionally used for the removal of waste when forming curves consisting of a U-shaped, sharpened knife blade with handles at either end. *See also* Adze.

Duodecagon A polygon having twelve sides.

Dwang Solid or herringbone strutting to a floor or the noggins of a stud partition.

Dwelling A unit of living accommodation intended to provide people with shelter from the external environment and a place for storage of their goods and chattels (all their movable property) such as an apartment, bungalow, cottage, flat, house or maisonette.

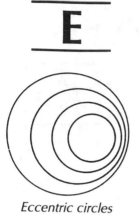

E

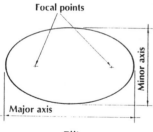

Eccentric circles

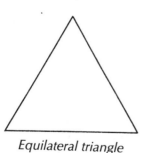

Focal points

Minor axis

Major axis

Ellipse

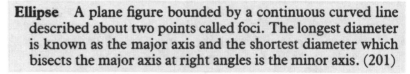

Equilateral triangle

Eccentric circles Circles that are drawn within each other from different centres.

Edge The perimeter of a surface or any line forming the intersection between two faces of a solid.

Elevation The view of an object from the side or front, rather than the plan which is from the top.

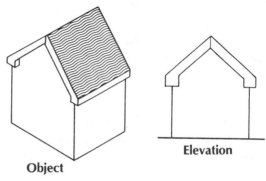

Object

Elevation

Elevation

Ellipse A plane figure bounded by a continuous curved line described about two points called foci. The longest diameter is known as the major axis and the shortest diameter which bisects the major axis at right angles is the minor axis. (201)

Environment Our physical surroundings such as people, buildings, structures, land, water, atmosphere, climate, sound, smell and taste. *See also* Built environment.

Equal Quantities which are the same. In mathematics the symbol = is used for equality. *See also* Mathematical symbols.

Equilateral triangle A triangle having all sides of equal length.

Eye The centre of a building element or component.

Fence Any construction that encloses land.

Fielded A panel with a raised centre part.

First angle A form of orthographic projection used for building drawings where in relation to the front view the other views are arranged as follows: the view from above is drawn below; the view from below is drawn above; the view from the left is drawn to the right; the view from the right is drawn to the left; the view from the rear is drawn to the extreme right. A sectional view may be drawn to the left or right. *See also* Third angle.

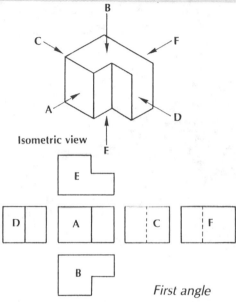

Isometric view

First angle

Flat A dwelling in a building of more than one storey, the building being subdivided horizontally into separate dwelling units, access to which is via a communal entrance. *See also* Apartment and Maisonette.

Floorboard saw A handsaw with a curved, rather than a straight, cutting edge used for crosscutting floorboards that are in position.

Formula A general mathematical rule, stated in algebraic form, such as LBH which is the formula for finding the volume of a rectangular prism where L = length, B = breadth and H = height. *See also* Area and Volume.

Fraction Part of a whole number, such as $\frac{3}{4}$ or 0.75 in decimals.

Frustum The remaining portion of a truncated solid figure. *See also* Section.

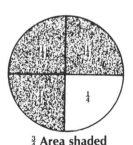

$\frac{3}{4}$ **Area shaded**

Fraction

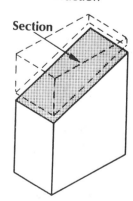

Frustum

G

Garden city An idea put forward by Ebenezer Howard in 1902 which suggested building medium-sized communities of about 32,000 people, each complete with their own facilities, industries and a planned layout surrounded by open countryside. This led to the building of Letchworth in 1903 and Welwyn in 1920. Houses each with their own garden were grouped in tree lined, grass verged crescents, closes, avenues and small side roads. This ideal can be seen to various extents throughout many housing estates built since.

Garden city

Geometry The study of the special properties of solid and plane figures.

Gradient A measure of the shape or inclination of the surface normally expressed as a ratio, such as a gradient of 1:10. This means that for every ten units travelled horizontally the surface rises one unit vertically. Also termed pitch.

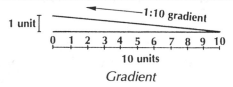

Gradient

Graph A diagrammatic representation of the relationship between two or more quantities.

Green belt Areas of open land around urban areas to prevent further expansion. They are kept open by severe and normally permanent planning restrictions.

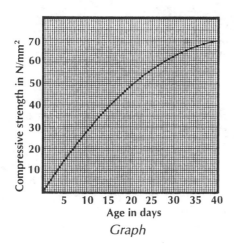

Graph

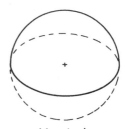

Hemisphere

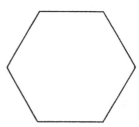

Hexagon

Helix   A spiral curved line, it lies on the surface of a cylinder or cone and cuts the surface at a constant angle.

Hemisphere   Half a sphere.

Heptagon   A polygon having seven sides.

Hexagon   A polygon having six sides.

High-rise   A building of over seven storeys in height. Sometimes termed a skyscraper. *See also* Low- and Medium-rise.

Horizontal   A plane or line laying from side to side, parallel with the skyline as opposed to vertical which is up and down, at right angles to the horizontal.

House   A building of two or more storeys, that is not subdivided vertically and is used as a dwelling.

Hypotenuse   The longest side of a right-angled triangle.

High rise

Horizontal

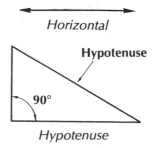

Hypotenuse

Index   A mathematical power.

Interest   A charge made for borrowing money.

Intersection   The point where two lines cross.

Inverse   The opposite of. Minus is the inverse of plus, division is the inverse of multiplication, etc.

Isometric   A pictorial method of drawing where all vertical lines of an object are drawn vertical and all horizontal lines are drawn at an angle of thirty degrees to the horizontal. *See also* Axonometric and Oblique.

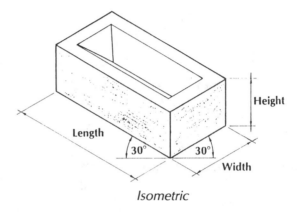

Isometric

Isosceles triangle

Isosceles triangle   A triangle having any two sides of equal length.

Japanese saw   A saw which cuts on the backward stroke rather than the normal forward stroke.

Jenny   A pulley or gin wheel used with a rope for hoisting small loads such as buckets or mortar to the upper floors of a building under construction. Also a set of callipers, one leg of which curves inwards while the other curves outwards.

Logarithm   A form of mathematics used before the introduction of calculators to simplify addition, subtraction, multiplication and division. The logarithm of a number to any given base is the power by which the base must be raised to give that number e.g. to a base 10

$$\log 100 = 2 \text{ as } 10^2 = 100$$

The inverse or opposite of a logarithm is an antilogarithm.

Maisonette   A dwelling occupying part of a larger building which is subdivided horizontally, but unlike a flat has its own separate outside entrance.

Major   The bigger portion. *See also* Minor.

Major axis   The longest diameter of an ellipse. *See also* Minor axis. (208)

Mansion   *See* heading under Architectural style.

Mathematical signs   A range of signs and symbols used as an alternative to writing out in full.

Sign	Meaning	Example
=	Equals	$4 = 3 + 1$
$\simeq$	Approximately equal to	$4 \cdot 9 \simeq 5$
+	Add	$4 + 5 = 9$
−	Subtract	$9 - 4 = 5$
×	Multiply	$4 \times 5 = 20$
÷	Divide	$20 \div 4 = 5$
$\sqrt{\ }$	Square root	$\sqrt{16} = 4$
n	Power	$4^2 = 16$

Maul   A large, long-handled mallet used when laying paving stones or flags.

Maximum   The most. *See also* Minimum.

Mean   Normally termed average. The mean value of several quantities is found by adding the quantities together and dividing by the number of quantities. For example the mean of 8, 10, 16, 20 and 27 is:

$$\frac{8 + 10 + 16 + 20 + 27}{5} = 16.2$$

See also Median and Mode.

Measurement *See* SI units.

Median The middle number of a set of figures which are arranged in ascending order e.g. the median of 8, 10, 16, 20 and 27 is 16. Where there is no middle number the median is taken as the mean of the two middle numbers e.g. the median of 8, 10, 16, 20, 27 and 34 is 18 since this is the mean of 16 and 20. *See also* Mode.

Metric *See* SI units.

Mews *See* heading under Architectural style.

Minimum The least. *See also* Maximum.

Minor The smaller portion. *See also* Major.

Minor axis The shortest diameter of an ellipse, it bisects the major axis at right angles. (208)

Minute A unit of angular measurement. Sixty minutes equals one degree. Also a unit of time, sixty minutes equals one hour.

Mode The number in a set of figures that occurs most frequently e.g. the mode of 2, 4, 3, 5, 3, 4, 7, 6, 7, 3 and 2 is 3. *See also* Mean and Median.

Multiplication One of the four basic mathematical operations. Concerned with repeated addition e.g. $6 \times 4 = 24$. This is the same as $6 + 6 + 6 + 6$. *See also* Addition, Division and Subtraction.

N

Nail set A nail punch.

Network An interconnected system of the sequence of operations.

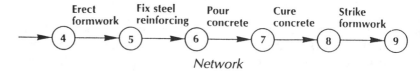

Network

Nonagon A polygon having nine sides.

Normal A straight line that starts at the centre of a circle and extends beyond the circumference. *See also* Tangent.

Oblique A pictorial method of projection termed either cabinet or cavalier where all vertical lines are drawn vertical and all horizontal lines in the front elevation are drawn horizontal to give a true front elevation. All horizontal lines in the side view are drawn at forty-five degrees to the horizontal. In cavalier these forty-five degree lines are drawn to their full length, but in cabinet they are only drawn to half length.

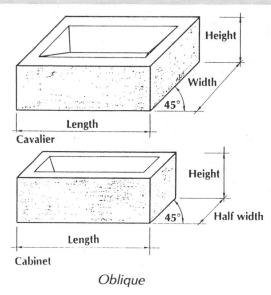

Oblique

Oblique solid A solid figure having an inclined central axis as opposed to a right solid.

Obsolescence Items or buildings, that become out of date or practice and fall into disuse. *See also* Derelict and Renovation.

Obtuse angle An angle greater than ninety degrees.

Obtuse-angled triangle A triangle in which one of the angles measures over ninety degrees.

Octagon A polygon having eight sides.

Orthographic A method of drawing plans and sections. A separate drawing of all the views of an object is produced in a systematic manner on the same drawing sheet. The two main methods of orthographic projection are known as first-angle projection which is used for building drawings and third-angle projection which is used for engineering drawings.

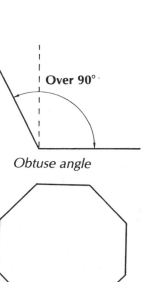

Obtuse angle

Octagon

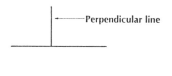

Parallelogram

Perimeter = 15,000 mm or 15 m

Perimeter

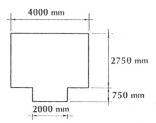

Perpendicular

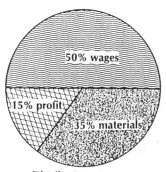

Distribution of earnings

Pie chart

Parallel   Lines that are positioned an equal distance apart. They will never meet no matter how far they are extended.

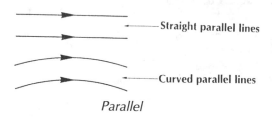

Parallel

Parallelogram   A quadrilateral with opposite sides of equal length but with no angle which is a right angle. (201)

Pentagon   A polygon having five sides.

Penthouse   *See* heading under Architectural style.

Percentage   A means of representing part of a quantity e.g. a fraction. Percentage means per hundred e.g. 25 per cent means twenty-five parts per hundred, $\frac{25}{100}$ or 0.25.

Perimeter   The distance or length around the boundary of a figure. This is a linear measurement given in metres run.

Perpendicular   At right angles to. Two lines are perpendicular if they meet at a right angle, hence a vertical line is perpendicular to a horizontal one.

Pi   A symbol π used to denote the number of times that the diameter of a circle will fit around its circumference. Normally taken to be 3.142, Pi is used when calculating lengths, areas and volumes of curved lines, figures and solids.

Pictorial projection   A method of drawing objects in a three-dimensional form. Often used for design and marketing purposes, as the finished appearance of the object can be more readily appreciated. This includes axonometric, isometric and oblique projection. *See also* Orthographic projection.

Pie chart   A circular diagram divided into sectors used to represent statistical information.

Plan   The shape of a surface or object when looked down on vertically. *See also* Elevation.

Plane   A flat surface.

Pollution Any direct or indirect alteration to the environment which is hazardous, or potentially hazardous, to health, safety and welfare of any living species.

Polygon A plane figure that is bounded by more than four straight lines and may be classified as either regular – having sides of the same length and equal angles or irregular – having sides of differing length and unequal angles. Also further classified by their number of sides. *See also* Pentagon, hexagon, heptagon, octagon, nonagon, decagon, undecagon and duodecagon.

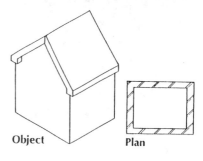

Object   Plan

Plan

Power A simple means of writing repeated multiplication also termed an index e.g. 6 to the power 2 (normally written at 6^2) = $6 \times 6 = 36$ and 6 to the power 3 (normally written as $6^3 = 6 \times 6 \times 6 = 216$. Numbers that are raised to the power 2 are said to be squared and numbers raised to the power 3 are said to be cubed. *See also* Roots.

Preservation The keeping in existence, unchanged, natural resources and buildings which have been inherited from the past. *See also* Conservation.

Prism A solid figure contained by plane surfaces which are parallel to each other. Named according to their base shape e.g. triangular prism, square prism and octagonal prism etc. Also classified as either right solids or oblique solids according to the inclination of their axes.

Product The result of multiplication.

Proportion *See* Ratio.

Protractor A device used to draw and measure angles.

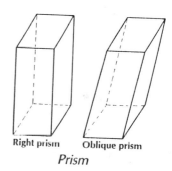

Right prism   Oblique prism

Prism

Purpose groups The division of buildings according to their main use. Either residential such as dwelling house, flat, institutional and other residential or non-residential such as assembly, office, shop, industrial and other non-residential.

Pyramid A solid figure contained by a base and triangular sloping sides. Named according to their base shape e.g. triangular pyramid, square pyramid, and octagonal pyramid etc. Also classified as either right solids or oblique solids according to the inclination of their axes.

Pythagoras's theorem A theorem for right-angled triangles which states that in any right-angled triangle the square of the hypotenuse is equal to the sum of the square of the other

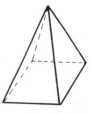

Pyramid

two sides. Thus a triangle having sides of three units, four units and five units must be a right angle since:

$$5^2 = 3^2 + 4^2$$
$$5 \times 5 = (3 \times 3) + (4 \times 4)$$
$$25 = 9 + 16$$
$$25 = 25$$

(See 225)

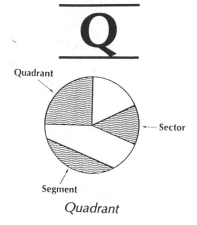

Quadrant

Q

Quadrant   A quarter of a circle. A sector having a right angle between its two radii.

Quadrilateral   A plane figure which is bounded by four straight lines. *See also* Square, Rectangle, Rhombus, Parallelogram, Trapezium and Trapezoid.

Quotient   The result of division.

R

Radii   Plural of radius.

Radius   Half the length of a circle's diameter. The distance from the centre to the circumference.

Ratio   Also termed proportion. Ratios are ways of comparing or stating the relationship between two like quantities e.g. £9 is to be shared between two people at a ratio of 1:2. There are three parts so one will receive £3 and the other £6.

Reciprocal   Any number divided into one e.g. the reciprocal of 25 is $\frac{1}{25}$ or 0.04.

Recreational accommodation   All buildings fulfilling a recreational role such as sports centres, swimming pools, stadiums, theatres, cinemas, concert halls, art galleries, libraries, and any assembly/meeting hall.

Rectangle   A quadrilateral having opposite sides of equal length and containing four right angles. (201)

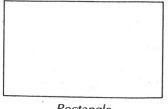

Rectangle

Rectilinear   Having, or along, a straight line.

Redevelopment   To demolish the existing buildings in an area, replan it and then rebuild.

Reflex   An angle greater than 180 degrees but less than 360 degrees.

Refurbishment   To bring an existing building up to standard or make it suitable for a new use by renovation.

Regular   Having equal length sides or angles, thus it is applied to polygons.

Rehabilitation   Slum areas and buildings brought up to an acceptable living standard by refurbishment.

Religious accommodation   Any building used as a place of worship to cater for a person's spiritual needs, such as a church, temple, mosque or synagogue.

Renovation   Bringing an existing building back to its former or original condition. Also termed restoration.

Residential   Buildings where people sleep on the premises such as dwellings, hotels, boarding houses, hospitals etc.

Restoration   The same as renovation.

Rhombus   A quadrilateral which has four equal-length sides which are parallel but has no angle which is a right angle.

Right angle   The angle formed when two lines touch at ninety degrees.

Right-angle triangle   A triangle in which one angle measures ninety degrees, the longest side of which is called the hypotenuse.

Right solid   A solid figure having a vertical central axis, as opposed to an oblique solid.

Root   The inverse of a power termed square root, cube root etc. The square root is a number multiplied by itself to give the number in question e.g. the square root of 25 (normally written as $\sqrt{25}$) is 5, since $5 \times 5 = 25$. The cube root of 125 (normally written as $^3\sqrt{25}$) is 5, since $5 \times 5 \times 5 = 125$.

Rural   In the country rather than an urban area.

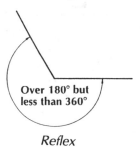

Reflex

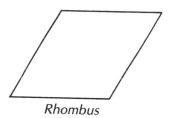

Rhombus

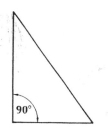

Right angled triangle

S

Scale   A series of marks used for measuring purposes. A scale drawing is a drawing produced using a ratio e.g. 1:5 which means that 500 mm full size is drawn as 100 mm.

Scale of chords   A rule used to set out angles from a radius and chord length.

Scalene triangle   A triangle having all three sides of unequal length.

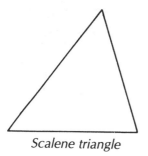

Scalene triangle

Section   The cut surface produced when a solid figure is cut through or truncated. *See also* Frustum. (209)

Sector   A portion of a circle that is contained between two radii and an arc. (201, 218)

Segment   A portion of a circle contained between an arc and a chord. (218)

Semi   Half of. For example a semicircle is half of a circle.

Semicircle   Half a circle formed by a chord passing through the centre point.

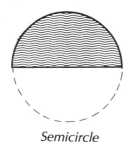

Semicircle

Semidetached   A house which is joined to one adjacent house, but is detached from all other buildings, thus it shares one party wall. *See also* Detached and Terraced.

Shop   A unit of working accommodation. A building where goods are sold including all places where an exchange of services takes place.

Semidetached

SI units (Système Internationale d'Unités)   A metric system of measurement. For each quantity of measurement there is a base unit, a multiple unit and a submultiple unit e.g. the base unit of length is the metre (m), its multiple unit the kilometre (km) is a thousand times larger and its submultiple unit the millimetre (mm) is a thousand times smaller: m $\times$ 1000 = km, m $\div$ 1000 = mm.

Side   A straight line or plane surface that forms a boundary of an object.

Sine   A trigonometric function.

Slant   The inclined distance between the vertex and a point on the base of pyramids and cones. Known as slant height.

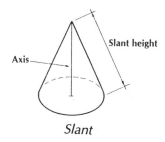

Slant

Slide rule   A calculating device used by engineers before the introduction of calculators.

Slum   An overcrowded, dirty, neglected, unhygienic area or building normally inhabited by poor people. *See also* Squat.

Solid figure   A three-dimensional object having length, width and thickness.

Solution   Finding an answer, especially in mathematics. *See also* heading under Materials and scientific principles.

Sphere   A solid figure described by the revolution of a semicircle about its diameter which remains fixed and is called the axis.

Square   Having an angle or corner of ninety degrees. Also used to define a quadrilateral having four equal-length sides and containing four right angles.

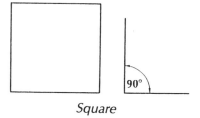

Square

Squat   To use a building or land without having ownership or permission from the owner. Squatting is often carried out as a means of free accommodation in slum areas which are due for rehabilitation.

Standard deviation   A measure of the spread of individual items of data from the mean.

Standard form   A method of expressing large or small numbers using an index or power e.g. 30,000 = 3 $\times$ 10,000 or 3×10^4; 0.036 = 36 $\times 10^{-3}$. The index is the number of places that a decimal point will have to be moved if the number is written in full. With a positive index the move is to the right and with a negative index the move is to the left.

Statistics The methods or the study of methods used in analysing large quantities of data.

Step One unit of a stair, consisting of one tread and one riser. Also the vertical distance between two horizontal surfaces.

Storage accommodation A building or space within a building used for storage or distribution of supplies.

Straight line The shortest distance between two fixed points.

Studio Traditionally an artist's workshop but now used for a one-room apartment or flat.

Subtraction One of the four basic mathematical operations e.g. $10 - 4 = 6$. It is the inverse of addition. *See also* Division and Multiplication.

Suburb A residential area situated on the outskirts of a town or city, hence suburban and suburbia. *See also* Urban.

Sum The result of addition.

Supplementary angles Two adjacent angles that make up 180 degrees. One is said to be the supplement of the other. *See also* Complementary angles.

Symbol A graphical representation used in drawing.

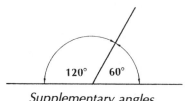

Supplementary angles

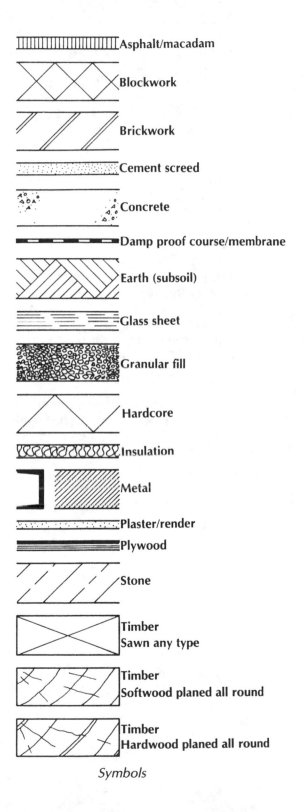

Asphalt/macadam

Blockwork

Brickwork

Cement screed

Concrete

Damp proof course/membrane

Earth (subsoil)

Glass sheet

Granular fill

Hardcore

Insulation

Metal

Plaster/render

Plywood

Stone

Timber
Sawn any type

Timber
Softwood planed all round

Timber
Hardwood planed all round

Symbols

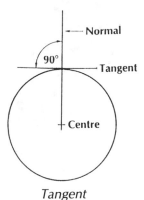

Tangent

Tenement

Tangent   A straight line that touches the circumference of a circle at right angles to the normal. Also a trigonometrical function.

Tell tale   A device, often glass, fixed across a crack in a wall to determine whether or not it is getting worse. Breakage of the device will indicate further movement.

Tenement   A block of apartments or flats, often with certain shared communal facilities such as washing, cooking or lavatories. Mainly situated in the poorer inner city areas.

Terrace   A row of three or more houses, the inner ones of which share two party walls. *See also* heading under Architectural style.

Theorem   A general conclusion. *See* Pythagoras's Theorem.

Third angle   An orthographic method of projection in which, in relation to the front elevation, the other views are arranged as follows: the view from above is drawn above; the view from below is drawn below; the view from the left is drawn left; the view from the right is drawn right; the view from the rear is drawn to the extreme right. Sectional views may be drawn on either the left or the right. *See also* First angle.

345 rule   A triangle with sides measuring three units, four units and five units respectively must be a right-angle triangle. Thus it is often used to set out square corners. *See also* Pythagoras's Theorem.

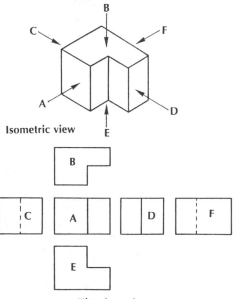

Third angle

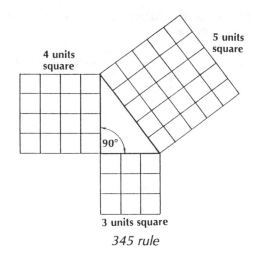

4 units square

5 units square

90°

3 units square

345 rule

Tonne   A unit of mass. One tonne = 1000 kg.

Town   A large collection of dwellings bigger than a village but not created a city.

Town house   Traditionally a well-to-do person's house in the town, owned in addition to their house in the country. Now taken to be any house over two storeys in height, often terraced. So-named because land in the town is more expensive resulting in town houses developing upwards rather than outwards as they would in the country.

Transposition   The rearrangement of a formula before carrying out a calculation. Basically anything can be moved from one side of the equals sign to the other by changing its symbol. This means that on crossing the equals sign the inverse is used. Plus changes to minus, multiplication changes to division, powers change to roots and vice versa. For example the formula for the circumference of a circle is: circumference = pi × diameter. This can be transposed to circumference ÷ pi = diameter.

Trapezium   A quadrilateral having two parallel sides. (201)

Trapezoid   A quadrilateral having no parallel sides.

Triangle   A plane figure that is bounded by three straight lines. *See also* Equilateral triangle, Isosceles triangle, Scalene triangle, Right-angled triangle, Acute-angled triangle, Obtuse-angled triangle. (201)

Trigonometry   A branch of mathematics that is concerned with the measurement of triangles. Angles and lengths of

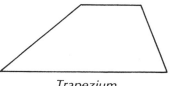

Trapezium

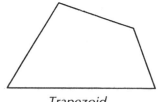

Trapezoid

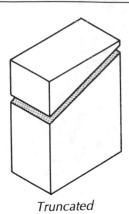

Truncated

sides can be calculated using the trigonometrical ratios of sine, cosine and tangent.

Truncated solid A solid figure with its top cut off. *See also* Section and Frustum.

Undecagon A polygon having eleven sides.

Units *See* SI units.

Unity The number one.

Urban A town or city area as opposed to a rural one.

Urbanization The process of change from a rural to an urban area.

Vertical

Vertex The highest point of intersection between adjacent sides of a pyramid.

Vertical A line at right angles to the skyline. It is perpendicular to the horizon or horizontal.

Vertices Points of intersection or corners.

Villa *See* heading under Architectural style.

Village A small collection of dwellings with a church in a rural area as opposed to a hamlet which is without a church.

Volume The space that an object takes up. It is a cubic measurement given in cubic metres (m^3) e.g. the volume of a cylinder is found by multiplying its base area by vertical height: volume = base area × height.

Working accommodation   The accommodation required for educational and industrial activities, offices and shops.

Workmate   A collapsible/portable vice and workbench used on site by carpenters. In many instances replaces the traditional carpenter's stool.

Zoning   A system of planning land use based on boundaries, inside which land can only be used for specified purposes such as agriculture, dwellings, green belt, industry or recreation etc.

5
MATERIALS AND
SCIENTIFIC
PRINCIPLES

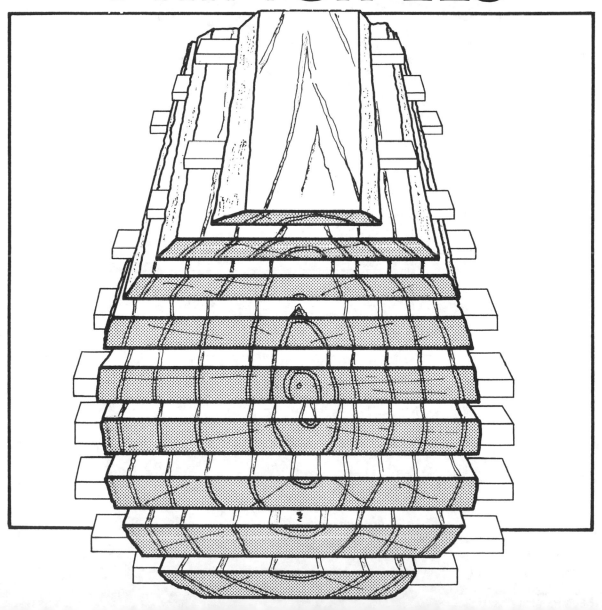

Abrasive   A material used for rubbing, grinding down or smoothing a surface such as emery cloth or paper, a grinding wheel and glasspaper etc.

Absolute zero   The lowest temperature that is theoretically possible. Zero degrees kelvin or −273 degrees celcius.

Absorbency   The ability of a substance to take in something else, normally water; or to reduce the intensity of something else such as heat, sound and light. *See also* Insulation and Capillarity.

Absorbent   A substance that has absorbency.

Acceleration   The increase in velocity per unit time measured in m/s^2 metres per second per second. *See also* Accelerator.

Acceleration due to gravity   The acceleration of a body falling freely in a vacuum. Taken to be 9.81 m/s^2. Although it will vary slightly depending on the distance from the earth's centre. *See also* Weight.

Accelerator   A substance which increases the rate of a chemical reaction. *See also* Catalyst, Curing, Retarder.

Acetylene   A poisonous flammable gas that when burnt with oxygen gives a flame temperature of up to 3300 degrees centigrade. Thus used for welding termed oxy-acetylene.

Acid   A substance containing hydrogen which can be replaced by a metal to form a salt. *See also* Acidic.

Acidic   A substance having the properties of an acid e.g. sour taste, some are corrosive, most will change an indicator, has a pH value between one and six, can be neutralized by a base solution to form a salt.

Acoustics   The study of sound.

Acrylic   A group of plastics used mainly as transparent roof sheeting or glazing. May also be used for baths and basins.

Action   In mechanics the effect produced by a force. The force of a box of nails sitting on the ground has an action on the ground. The ground does not move but pushes back with an equal reaction in the opposite direction. *See also* Equilibrium.

Adhesion   The sticking together of materials by chemical or mechanical bonds e.g. the force of attraction between molecules of different substances termed chemical adhesion; the penetration of an adhesive into the porous surface of a material forming mechanical bonds. *See also* Cohesion.

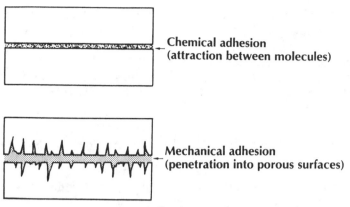

Adhesion

Adhesive A substance possessing the power of adhesion. Sticks surfaces together by either chemical attraction or mechanical bonds. May be classified as either natural or synthetic.

Admixture A material added to the basic constituents of a mix to alter one or more of its properties.

Affinity The force that binds atoms together in molecules; a chemical attraction.

Aggregate A filler material used in mortar and concrete mixes. Variously named according to their source or particle size. *See also* Fine, Coarse, All-in, Lightweight, Natural and Manufactured aggregate.

Aggregate cement ratio The relationship between the amounts of aggregate and cement in a concrete mix.

Air The atmosphere; the gas that surrounds the earth.

Air drying The drying of materials or coatings in natural conditions or at normal room temperatures as opposed to artificial or forced drying which requires heat. *See also* Artificial seasoning and Natural seasoning.

Air seasoning The natural drying or seasoning of timber in the air.

Air set The partial hydration of cement and plaster prior to use, due to the absorption of water vapour from the air. Also termed bag set.

Airborne sound   *See* Sound insulation.

Alabaster   A pure fine form of gypsum.

Alkali   A soluble base. *See also* Alkaline.

Alkaline   Having the properties of an alkali, the opposite of acidic. Will neutralize an acid in solution to form a salt, changes an indicator, has a pH value of between eight and fourteen, may be caustic.

All-in aggregate   A graded mixture of fine and coarse aggregates used in concrete. Also termed ballast.

Alloy   A mixture of two or more metals such as brass from zinc and copper, bronze from copper and tin, aluminium alloy from aluminium and magnesium. *See also* Ferrous and Non-ferrous metals.

Alternating current (a.c.)   An electric current produced by a generation in which the electron flow regularly changes direction. *See also* Direct current.

Altitude   Height.

Aluminium   A lightweight non-ferrous metal extracted from bauxite by electrolysis. May also be mixed with magnesium to form a stronger aluminium alloy.

Ammeter   A device for measuring electric current.

Amorphous   Non-crystalline having no definite shape or form.

Ampere (amp)   A unit of electric current. The amount of current flowing through a conductor. Amps $= \dfrac{\text{watts}}{\text{volts}}$.

Amplitude   The distance between the middle and the top of a sound or electromagnetic wave.

Anhydrite   The naturally occurring form of calcium sulphate.

Anhydrous   Applied to a substance containing no water. The opposite of hydrated and hydrous. Also a type of gypsum plaster used for finishing. *See also* Hemihydrate.

Animal glue   An adhesive derived from animal bones, hooves or skin.

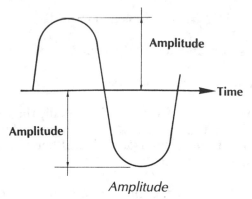

Amplitude

Anion   A negative charged ion that is attracted to the anode. *See also* Cathode and Cation.

Annealing   The slow, regulated cooling of metals, carried out to relieve stresses caused by working or other treatment.

Annual ring   One year's growth of a tree. One of the rings seen on the end grain of timber consisting of a wide band of lighter spring growth or early wood and a narrow band of darker summer growth or late wood. *See also* Hardwood and Softwood.

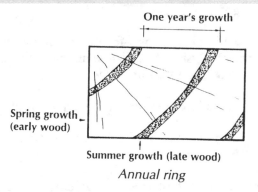

Annual ring

Annular   Ringed or having rings.

Annular nail   A wire nail with a ringed or serrated shank, which increases its holding power.

Anode   The electrode that is connected to the positive terminal of an electric supply. In electrolytic corrosion the metal that decomposes. *See also* Cathode.

Anodizing   The forming of a protective oxide coating on aluminium by electrolysis. May be a glossy or mat finish.

Anti-capillary groove   A groove cut into one or both adjoining faces of external joinery components to create a large gap, breaking the effect of capillary attraction in the otherwise close joint. Mostly used in the joint between jambs and stiles of windows and doors.

Aqueous A watery solution.

Arc The flash of light accompanied by intense heat produced when an electric current flows through a gap between two electrodes, as in arc welding. *See also* heading under General.

Archimedes' Principle The apparent loss of weight of an object when it is immersed, wholly or partly, in a liquid caused by an upthrust. The upthrust is equal to the weight of the liquid displaced. *See also* Buoyancy.

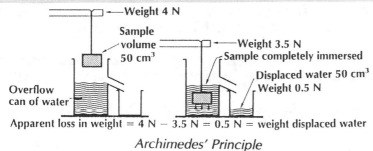

Archimedes' Principle

Architectural ironmongery High-quality or specially-designed door and window furniture.

Artificial seasoning The seasoning or drying of timber by artificial means; normally by kiln seasoning but chemicals, press drying, microwave and dehumidifying equipment may also be used.

Atmosphere The gases around the earth. *See also* Atmospheric pressure.

Atmospheric pressure The pressure on the earth's surface caused by the atmosphere. This varies from day to day according to the weather. Normal pressure is taken as 101,325 pascals, or approximately one bar.

Atom The smallest piece of an element which can take part in a chemical reaction. *See also* Electron and molecule.

Atomization The process of breaking up liquids, normally paint or varnish, into a spray of fine particles by compressed air.

Audibility The limits of frequency of sound. Human hearing responds to frequencies between 20 and 20,000 hertz. *See also* Decibels.

B

Balance A device used for weighing. *See also* Weight and Equilibrium.

Ballast All-in aggregate.

Bar A unit of pressure equal to 100,000 N/m^2 (Newtons per square metre). One bar is approximately atmospheric pressure. *See also* Pascal.

Barometer A device for measuring atmospheric pressure.

Base A substance that reacts with an acid to form a salt and water only. *See also* Alkali.

Batching The process of proportioning the constituent materials for a concrete mix, either by weight or by volume.

Battery A group of two or more primary cells coupled together. These may be arranged either in series or parallel.

Bauxite Natural aluminium oxide. The key ore of aluminium.

Beech A hardwood from Europe and Japan used for furniture, especially the framework of upholstered seating, plywood and high-quality floor covering. Light brown in colour with golden flecks.

Bending moment The moment of a force that sets up a bending tendency in a beam. *See also* Maximum bending moment.

Bi A prefix used to denote two. *See also* Bi-metal strip.

Bi-metal strip A strip made up of two different metals fixed together when heated differential expansion causes the strip to buckle. Used in thermostats.

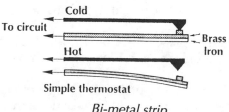

Bi-metal strip

Bitumen A thick liquid or solid that will soften and flow on heating. It occurs either naturally or may be distilled from petroleum. Being impervious it is used for mastic asphalt, adhesive, paints, roofing felt and damp proof courses.

Bituminous   Containing bitumen or tar.

Blast furnace   A furnace used for the smelting of iron from ore.

Bleach   A liquid used to remove or reduce the colour of materials by chemically changing the dye into a colourless substance.

Bleeding   Excess water rising to the surface of freshly-placed concrete. This leaves behind a network of interconnected voids which reduces both the strength and the durability of the concrete. *See also* heading under Services and finishes.

Block   A walling unit which is larger than a brick. Normally either concrete or natural stone.

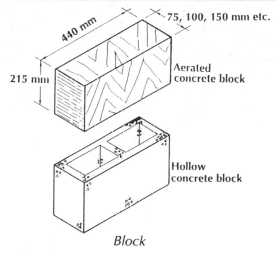

Block

Blockboard   *See* Laminated board.

Blowhole   A small hole or cavity in the concrete face due to air pockets trapped against the form faces; or excessive application of release agents; or use of neat oil as a release agent.

Boiling point   The temperature at which bubbles of vapour are starting to form rapidly in a liquid. For water this is 100 degrees centigrade.

Bond   The force that holds together a group of atoms or ions. *See also* heading under Building construction.

Bonding agent   An adhesive material normally applied to a smooth surface to increase adhesion.

Bowing   A timber distortion, a curvature along the face of a board.

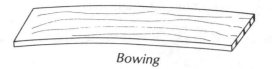

Bowing

Brass   A non-ferrous metal or alloy formed by a combination of zinc and copper.

Breather membrane   A building paper which is moisture-resistant but permeable to water vapour. Used as a moisture barrier under cladding.

Bricks   A walling unit component having a standard format size including a 10 mm mortar allowance of 225 mm × 112.5 mm × 75 mm. Two main categories are calcium silicate or clay. Clay bricks are usually pressed, cut or moulded and then fired in a kiln at very high temperatures. Their density, strength, colour and surface texture will depend on the variety of clay used and the temperature. Calcium silicate bricks are pressed into shape and steamed at high temperature. Pigments may be added during the manufacturing process to achieve a range of colours. The three main types of bricks are commons or flettons, facing bricks and engineering bricks.

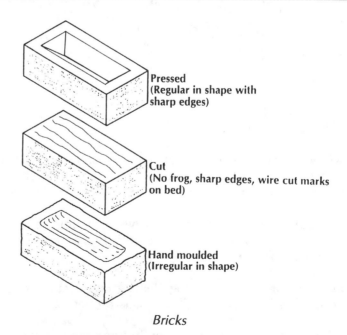

Pressed
(Regular in shape with sharp edges)

Cut
(No frog, sharp edges, wire cut marks on bed)

Hand moulded
(Irregular in shape)

Bricks

Broad leaf   A hardwood tree.

Bronze   A non-ferrous metal or alloy formed by a combination of copper and tin.

Building paper   A waterproof paper used under claddings. Formed from a fibre-reinforced bitumen layer sandwiched between sheets of brown craft paper. *See also* Breather membrane.

Bulking   The increase in the volume of damp aggregate due to the film of liquid separating the particles.

Dry aggregate particle   Damp aggregate particle   – Film of moisture

Bulking

Buoyancy   The tendency of an object to float in a liquid. The upward thrust of a fluid on an immersed object. *See also* Archimedes' Principle and relative density.

Burning   *See* Combustion.

Butane   A liquefied petroleum gas used as a fuel.

Butterfly wall tie   A wall tie used for cavity walls made from copper or galvanized steel wire, twisted to form two triangles with touching apexes.

By-product   A substance which is incidentally obtained during the manufacture of another substance.

Calcination   The extreme heating of a substance, thus used during the manufacture of cement, lime and plaster.

Calcium   A soft white metal that tarnishes rapidly in the atmosphere. Its main compounds are calcium carbonate, calcium hydroxide, calcium silicate and calcium sulphate.

Calcium carbonate   Occurs naturally as chalk, limestone and marble. Used in the manufacture of lime and cement.

Calcium hydroxide   Slaked lime used in cement, plaster and mortar.

Calcium silicate   A clear crystalline substance used in the manufacture of bricks, cement and glass.

Calcium sulphate   Anhydrite. Also gypsum. May be heated to form plaster of Paris.

Calibration   The graduations of an instrument which enable accurate measurements to be made.

Calorie   The unit of quantity of heat. The amount of heat required to raise the temperature of one gramme of water one degree Celsius. One calorie is equal to 4.2 joules.

Calorific value   Applied to a fuel it is the quantity of heat produced by a given mass of fuel on complete combustion. Expressed in J/kg (joules per kilogramme). Also known as heat of combustion.

Candela   A unit of luminous intensity.

Capacitance   The property of an electrical system which enables it to store an electric charge.

Capacity   The volume of a container measured in litres. Also the output of an electrical apparatus.

Capillary attraction or capillarity   The phenomenon whereby a liquid can travel against the force of gravity, even vertically in fine spaces or between surfaces close together, due to its own surface tension. The smaller the space the greater the attraction. *See also* Anti-capillary groove.

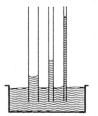

Liquid travels highest in narrow tube by capillarity

Capillary attraction

Carbon dioxide   A gas heavier than air that does not support combustion, used in fire extinguishers and as a refrigerant. Also used by plants during photosynthesis.

Case hardening   A timber defect caused by too rapid kiln drying. Often not apparent until a piece is re-sawn where it will tend to twist or bind on the saw.

Casein   An adhesive used for timber, derived from soured, skimmed milk curds, which are dried and crushed into a powder. Mixed with water to use, has gap filling properties, but will stain hardwoods.

Cast iron   A ferrous metal. A brittle form of iron containing impurities. Also known as pig iron.

Catalysis   *See* Catalyst.

Catalyst   A substance that accelerates a chemical reaction but does not get used up itself. The process is termed catalysis.

Cathode   The electrode that is connected to the negative terminal of an electric supply. In electrolytic corrosion the metal that is given sacrificial protection. *See also* Anode.

Cation   The positive ion that is attracted to the cathode in galvanic corrosion. The metal that does not decompose. *See also* Anode and Anion.

Cavitation   A phenomenon associated with water flow, consisting of the formation and collapse of cavities in the water. Results in premature wear to valves and seals in plumbing systems.

Cell   The smallest part of a plant or animal such as parenchyma, tracheid and pores which are timber cells. *See also* Cellulose.

Cellular   A brick where depressions or cavities exceed 20 per cent of its gross volume.

Cellulose   A fibrous structure which forms the cell walls of plants.

Celsius   A temperature identical to centigrade measured on a scale having 100 degrees between 0 degrees (the melting point of ice) and 100 degrees (the boiling point of water). *See also* Kelvin.

Cement   A fine powder which, when mixed with water, forms a paste that gradually hardens. Acts as an adhesive in concrete, bonding aggregate together. *See also* OPC, RHPC, SRPC, HAC, cement fondue and hydration.

Cement fondue   A rapid hardening cement made from bauxite and lime, rather than Portland based. *See also* HAC.

Cement-sand plaster   Used for external rendering, internal undercoats and water-resisting finishing coats.

Centigrade   *See* Celsius.

Centre of gravity   A fixed point of a body about which all its parts balance, irrespective of the body's position. Its weight is considered to be acting through its centre of gravity. In simple objects the centre of gravity is its geometric centre.

Centrifugal force   The outward force that acts on a body rotating in a circle around a fixed point.

Ceramic   Burnt clayware, consisting of a mixture of sand and clay, shaped, dried and finally fired in a kiln. Main types and uses are: terracotta – mainly for unglazed air bricks, chimney pots and floor tiles; faïence – a glazed form of terracotta and stoneware; fireclay – has a fire resistance and is used for firebacks and flue linings; stoneware – more glassy than fireclay and is used for underground drainage goods; earthenware – normally white, has fine texture and a highly glazed surface and is used for wall tiles; vitreous china – more glassy than earthenware and is used for sanitary appliances.

Change of state   The change of a material from one state of matter to another, such as from a solid to a liquid and a liquid to a gas. These changes are mainly the result of heating or cooling.

Charge   The property of particles. Electrons are negative, protons are positive. Like charges will repel each other and opposite charges will attract each other.

Chemical   Any substance used in or produced by chemistry. *See also* Chemical change, Formula and Reaction.

Chemical change   A change in a substance in which a new substance is formed with different properties. Normally accompanied by heat. *See also* Chemical reaction.

Chemical formula   A form of shorthand for a substance which shows the number and types of atoms in a molecule e.g. water is H_2O which means two atoms of hydrogen and one atom of oxygen for each molecule of water.

Chemical reaction   A chemical change that takes place when two or more substances are put together. This results in a rearrangement of the bonding of atoms or ions and a new substance is formed.

Chipboard   *See* Particle board.

Chlorination   The process of adding chlorine to drinking water and swimming pools to kill harmful bacteria.

Chromium   A metal used with steel to form stainless steel. Also as chromium plating.

Circuit   A continuous path along which an electric current can flow.

Coal tar   A thick oily liquid derived from coal. Used as a timber preservative, with an aggregate in Tarmacadam and also for other bitumen applications.

Coarse aggregate   Having particles mainly greater than 5 mm in size.

Coefficient of thermal expansion   The increase in length (linear expansion) per unit length of a material, caused by a rise in temperature of one degree C or K. May also be applied to area (superficial expansion) or volume.

Cohesion   The strength within a substance. The force of attraction between molecules of the same substance. It is the force that holds solids and liquids together. *See also* Adhesion.

Colloidal   A dispersion of fine particles in a solvent, an intermediate state between a solution and a suspension.

Colombian pine   Douglas fir.

Colour   The effect on the eye produced by different wavelengths of visible light.

Combustible   A solid material that is capable of burning as opposed to a non-combustible one which is incapable of burning. *See also* Flammable.

Combustion   A chemical reaction, also called burning, between a substance and oxygen, during which heat is produced and the original form of the substance is destroyed. The essential requirements for combustion to take place are a combustible or flammable material to provide fuel; oxygen, normally from the air, to combine with the flammable vapour given off by the material and initial heat to bring the material up to its ignition temperature or flashpoint.

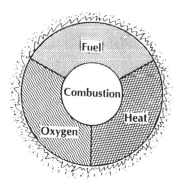

Combustion

Common or fletton bricks   These are not chosen for their appearance or strength. They are basic bricks used in the main, for internal or covered (rendered or cladding) external work, although sand-faced flettons are often used as a cheap facing brick.

Compound   Two or more elements that have been bonded together by a chemical reaction.

Compression   Stress in a structural member that causes squashing and crushing. It has a shorting effect, the opposite of tension. *See also* Shear.

Concentrated solution A solution that contains a large proportion of a dissolved substance. It is a strong solution. *See also* Dilute.

Concrete A composite material made from a mixture of cement, aggregate and water. Classed as either plain, reinforced, prestressed, in situ cast, precast or lightweight. *See also* Hydration and Curing.

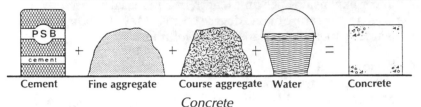

Cement   Fine aggregate   Course aggregate   Water   Concrete

Concrete

Condensation The change of a vapour or gas to a liquid state, at the dew point, such as the condensation of water vapour on a cold surface due to the fact that cooler air has a lower vapour-carrying capacity than warmer air. *See also* Surface and Interstitial condensation.

Condense A change of state from a vapour to a liquid.

Conditioning Second seasoning; the prior storage of joinery and trim in the area where they are to be fixed, so that an equilibrium moisture content can be achieved; also the practice of brushing water into the mesh side of hardboard about twenty-four hours before fixing so that the board shrinks on its fixing and does not expand, causing buckling.

Conduction The conduction of heat within a material from hotter to cooler areas. *See also* Convection and Radiation.

Conductivity The rate at which a material conducts heat. *See* Thermal conductivity. Also refers to the conductivity of an electrical conductor. A measure of resistance.

Conductor A material with a high thermal conductivity. It is a poor insulator, as opposed to a material with a low thermal conductivity which is a good insulator. Also a substance that allows electricity to pass through it.

Convection The transfer of heat within liquids and gases. The heated particles expand and become less dense; this causes them to rise, their place being taken by colder denser particles. *See also* Conduction and Radiation.

Convection currents Movement in liquids and gases when heated by a process of convection.

Conversion   The sawing up of a tree trunk into variously-sized pieces for a specific purpose. Also a change in the crystalline structure of a material. *See also* HAC.

Corrosion   A surface chemical reaction. Especially applied to metals which can be 'eaten away' or corroded by the action of water, air and chemicals. *See also* Oxidation.

Coulomb   The unit of electric charge. The quantity of charge transferred by one ampere in one second between any two points in a circuit e.g. a current of 10 amps flowing for 5 seconds transfers 50 coulombs.

Creosote   A tar-oil based timber preservative only suitable for external use and never in association with foodstuffs.

Crystal   A solid that has formed naturally into a characteristic shape.

Crystalization   The process in which crystals are formed.

Cube test   A test carried out on hardened concrete cubes to determine their crushing strength in N/mm^2, normally at twenty-eight days after casting.

Cupping   A timber distortion taking the form of a curvature across the width of a tangentially sawn board.

Cured   A substance that has completed the curing process.

Curing   The process of changing from a liquid to a solid by a chemical reaction. Applied to the hardening process of adhesives, cement, concrete, plastics and plasters etc. Also the process of retaining water in recently poured concrete and providing insulation from extremes of temperature. *See also* Accelerator, Hydration, Retarder, and Setting.

Current   The movement of electrons through a conductor measured in amperes.

Curtains   Hardened grout running down the face of completed concrete work. Often accompanied by grout loss or honeycombing caused by poor sealing between formwork and completed structure. Also excessive interconnected runs in paintwork.

Cycle   A series of operations or changes performed which return to the original state or position.

D

Decay The decomposition of timber brought about by a fungus; causes a breakdown of the cellular structure, softening and loss of strength. Mainly attributed to wet and dry rot fungi.

Decibel (dB) A unit used to measure the intensity of sound.

Deciduous A hardwood tree that loses its leaves in winter, as opposed to evergreen which replaces leaves throughout the year.

Deflection The amount of bending or sagging of a beam under load. Normally limited to a maximum of 3 mm in every metre run.

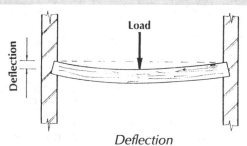

Deflection

Dehumidifier A device, hydroscopic material or deliquescent that removes or absorbs moisture vapour from the air in order to reduce humidity.

Dehydration The removal or elimination of water.

Deliquescent A substance that absorbs water from the air. *See also* Hydroscopic.

Dense A substance that is heavy for its volume.

Density The mass of a unit volume of a substance. Expressed in kilogrammes per cubic metre (kg/m^3). The density of a 1 m^3 of material having a mass of 750 kg is 750 kg/m^3 as:

$$\text{Density} = \frac{750 \ (\text{mass})}{1 \ (\text{volume})} = 750 \ \text{kg/m}^3$$

The smaller the space into which the mass is concentrated, the greater the density. *See also* Specific gravity.

Dew The liquid (water) produced when moisture vapour condenses.

Dew point The point at which air becomes saturated with water vapour, this is a relative humidity of 100 per cent. Further cooling causes condensation.

Differential expansion   The different coefficient of thermal expansion of various materials. Particularly refers to materials that are joined such as a bi-metal strip.

Diffuse porous   *See* Pores.

Diffusion   The spreading of the molecules of a substance through a solid, liquid or gas.

Dilute   A solution that contains a small amount of a dissolved substance. It is a weak solution.

Dilution   A dilute solution. *See also* Concentrated solution.

Direct current (d.c.)   An electric current that flows in one direction only. *See also* Alternating current.

Discharge   The loss of an electric charge.

Dispersion   Fine particles in a solvent not a solution or a suspension. *See* Colloidal.

Dissolved   A solid broken down in a liquid into such small particles that a clear solution is formed.

Distillation   The separation of a dissolved solid from its solvent.

Distortion   Out of true shape. In timber the bowing, cupping, springing or winding of a board.

Douglas fir   A softwood from Canada and USA also known as Colombian pine used for all general carpentry and joinery work, for both structural and non-structural applications; also plywood for sheathing and decking. Generally considered to be of better quality than European Redwood. Has a pale red/brown heartwood and lighter sapwood.

Dry cell   A battery containing no liquid.

Dry ice   Carbon dioxide in a solid state used for refrigeration.

Dry rot   A fungi that feeds on, and destroys, damp rather than wet timber. Most often found in damp, poorly ventilated underfloor spaces and roof spaces. Causes timber to lose strength and weight, develop cracks in brickbond patterns (cuboidal) and finally become so dry and powdery that it is easily crumbled. *See also* Wet rot.

Ductile   A metal that can easily be drawn out into a wire. *See also* Malleable.

Dynamics   The study of moving objects and the forces that cause changes in motion, rather than statics.

E

Earth   *See* heading under Services and finishes.

Earthenware   *See* Ceramics.

Earthing   *See* heading under Services and finishes.

Echo   The effect produced when sound is reflected on meeting a solid obstacle. *See also* Reverberation.

Efflorescence   Hydrated substances that lose water to the air. In building, white deposits on the surfaces of walls caused by soluble salts, which crystalize when the structure dries out.

Elastic limit   The limit of a material's elasticity, the point beyond which a material will not completely return to its original shape when the stress is removed.

Elasticity   The ability of a material to return to its original shape or size after being subjected to stress. *See also* Elastic limit.

Electric charge   Moving electricity for power, rather than static electricity.

Electric current   The flow of an electric charge through a conductor, measured in amperes.

Electric power   The rate of doing work measured in watts. *See also* Kilowatt hour

Watts = volts × amperes

Electro-chemical series   A list of metals arranged in order of decreasing reactivity. Those higher up the list will decompose in preference to those lower down during electrolytic corrosion. The main metals used in building, in descending order are zinc, cadmium, iron, nickel, tin, lead, copper and aluminium.

Electrode   A piece of metal used to carry an electric current either into or out of a solution or liquid.

Electrolysis Passing an electric current through a solution to bring about a chemical change.

Electrolyte A solution which will undergo electrolysis. *See also* Electrolytic corrosion.

Electrolytic corrosion Corrosion, also termed galvonic corrosion, that occurs when two different metals are in contact in the presence of an electrolyte. This forms a simple cell in which one of the materials is decomposed. *See also* Electrochemical series, Sacrificial protection, Anode and Cathode.

Electron Part of an atom having a negative charge. *See also* Proton and Neutron.

Element A pure substance that cannot be broken down by a chemical reaction into anything simpler. It consists of atoms. *See also* heading under Building construction.

Emulsion A water-thinned paint for use on walls and ceilings.

Endothermic A chemical reaction that takes in heat unlike exothermic which releases heat.

Energy The ability to do useful work. The unit of energy is the joule (J). *See also* Power.

Engineering bricks These have a very high density and strength and do not absorb moisture. They are thus ideal for use in damp conditions such as inspection chambers, basements and other below-ground work. Also used in highly loaded conditions such as the substructure of tall buildings or columns supporting a beam.

Equilibrium Balanced. A material's moisture content is in equilibrium when it matches the relative humidity. A structural member or lever etc. is in equilibrium when the sum of the clockwise moments equals the sum of the anticlockwise moments. *See also* Principle of moments. (271)

European redwood A common softwood also known as Scots pine. Used for all general carpentry and joinery work for structural and non-structural applications. Has a pale red/brown heartwood and a light yellow/brown sapwood.

Evaporate The loss of a liquid's surface molecules into the air by evaporation.

Evaporation   The change in state from a liquid to a gas below boiling point. *See also* Vapour.

Evergreen   A tree, normally softwood, that maintains its leaves throughout the year, as opposed to deciduous that loses its leaves in the winter.

Exothermic   A chemical reaction that releases heat e.g. combustion, unlike endothermic which takes in heat.

Expanded metal   A sheet metal with an open expanded mesh.

Expanded polystyrene   An air expanded plastic obtainable in blocks and sheets.

F

Facing brick   These are made from selected clays and chosen for their attractive appearance rather than any other characteristic.

Faïence   *See* Ceramic.

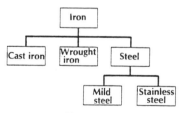

Ferrous

Ferrous metals   These are all extracted from iron ore with varying amounts of carbon added to them e.g. wrought iron, cast iron, mild steel, high tensile steel. They will corrode rapidly when exposed to air and water, until they are completely rusted. Chromium and nickel may be added to form a stainless steel alloy that is resistant to rusting.

Fibre   The main structural tissue of hardwoods, which gives its mechanical strength. *See also* Parenchyma.

Fibre saturation point   Timber at a moisture content of about 25 per cent to 30 per cent. Shrinkage will occur when the moisture content is lowered below fibre saturation point. (269)

Fibreboard   A sheet material made from pulped wood, mixed with an adhesive and pressed forming hardboard, medium-board, and insulation board (floor, wall, ceiling and formwork linings, insulation and display boards). *See also* Medium density fibreboard (MDF).

Fibreglass   A thermal insulation material consisting of glass fibre strands woven into sheets or rolls. *See also* Glass reinforced plastic.

Field settling test   An on-site test used to determine the percentage of silt present in a fine aggregate.

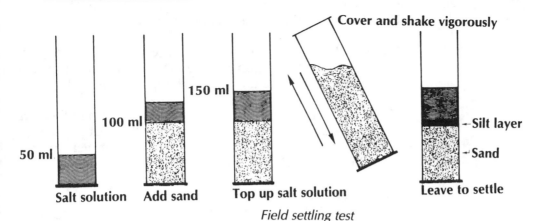

Field settling test

Filament   A piece of thin wire having a high resistance. When an electric current is passed through, it becomes hot and gives off light e.g. an electric lightbulb.

Filler   A solid substance added to another material in order to reduce the cost, or modify its properties. Used in adhesives, paints, plastics etc. e.g. aggregate is a filler in concrete.

Film   A thin layer of a substance formed on the surface of a liquid.

Filter   A device used to separate solids from liquids. *See also* Filtration.

Filtration   The process of separating solids from liquids by passing through a filter which holds back the solid but permits the liquid to trickle through.

Fine aggregate   Having particles mainly less than 5 mm in size e.g. sand.

Finishing coat   The paint coat that seals the surface, gives the final colour and provides the desired surface finish (flat, eggshell, gloss). Also the final coat of plaster.

Fire   Burning or combustion.

Fire resistance   A property of an element of building construction, such as a wall, roof or floor. The ability of an element to satisfy stated criteria when subjected to a standard fire test. Fire resistance is defined by reference to its stability, integrity and insulation ratings.

Fireclay   *See* Ceramic.

Fixings   Nails, screws and bolts etc. *See also* Ironmongery and Hardware.

Flamespread *See* Surface spread of flame.

Flammable A liquid or gas that is capable of burning as opposed to a non-flammable one which will not burn. *See also* Flashpoint, Combustible and Non-Combustible.

Flashover A stage in the development of a building fire where simultaneous ignition of the flammable vapours occurs causing a rapid rise in temperature and merged balls of flame to travel throughout the space.

Flashpoint The lowest temperature at which a liquid, material or gas will give off a flammable vapour. *See also* Ignition temperature.

Fletton A common clay brick originally from a village near Peterborough.

Flexibility The measure of a material's ability to bend or flex.

Float glass Flat glass that gives an undistorted vision made by floating molten glass onto the surface of liquid tin, and subsequently allowing it to cool.

Fluid Any material that flows – a liquid or a gas.

Fluorescence An ability of a substance to absorb energy and emit it at another wavelength. *See also* Fluorescent.

Fluorescent A substance that exhibits fluorescence. Fluorescent paints and fluorescent tubes contain chemicals which absorb ultra-violet light and release the energy as visible white light.

Flux A substance used in soldering which increases the fusibility of metals.

Force The push or pull which causes acceleration, a change in shape or a reaction. It is measured in newtons. A mass of 1 kg is said to exert a force due to gravity of 9.81 N. For practical purposes 1 kg is normally taken to exert a force of 10 N.

Freezing point The temperature at which a liquid solidifies. It is the same as the melting point e.g. for water 0°C.

Frequency In sound the number of vibrations or waves in one second. The unit of frequency is the hertz. Something vibrating at 500 times a second will have a frequency of 500 hertz.

Friable   Easily crumbled in the hand.

Frogged   A brick with a depression in one or both of its bed faces, which in total are up to 20 per cent of its gross volume.

Frost resistant   A clay brick classification (F) meaning that they are durable in all situations.

Fuel   A substance which releases heat energy, normally during combustion. *See also* Combustible and Flammable.

Fungi   Simple plants which can cause the decay of timber, such as dry rot and wet rot. *See also* Mould.

Fungicide   A sterilizing solution. *See also* Preservative.

Furniture   The locks and handles fixed to doors and windows. *See also* Hardware and Architectural ironmongery.

G

Galvanized   The coating of iron or steel with a thin layer of zinc by dipping it into molten zinc. The result is that the zinc offers sacrificial protection to the iron. *See also* Sherardizing.

Gas   A state of matter in which molecules are free to move about as they have few bonds between them. Gas has no definite volume and no shape. *See also* Solid and Liquid.

General structural or (GS)   A visual, stress-graded timber used for structural applications. Machine general structural (MGS) is the machine graded equivalent. *See also* Special structural.

Genus   The grouping together of trees having common characteristics e.g. *pinus* is the generic name for all pine softwoods such as Scots Pine and Pitch Pine etc.

Gilding   The application of gold leaf.

Glass   A hard, transparent mixture of silicates formed by melting together the basic constituents of sand, soda ash, limestone and dolomite. On cooling the molten glass hardens and becomes clear. Most flat glass is termed float glass.

Glass reinforced plastic   Often abbreviated to GRP. A material produced by the impregnation of woven, glass fibre cloth with a synthetic resin. Also termed glassfibre.

Glassfibre   Glass reinforced plastic. *See also* Fibreglass.

Glue   An adhesive, an animal glue.

Gold leaf   Thin sheets of gold available in book form about 100 mm square, used for gilding. *See also* Gold size.

Gold size   An oleo-resinous varnish of two types: used to fix gold leaf it quickly becomes tacky but hardens slowly; used as a filler it has a much higher proportion of driers to make it harden.

Grade stress   The stress that can be safely carried by a particular grade of structural timber. A modification of the basic stress to allow for the inclusion of strength-reducing defects. *See also* Permissible stress.

Grading   The selection or classification of materials for quality or strength or particle size. *See also* Sieve analysis and Stress grading.

Gramme (g)   A submultiple of a kilogramme.

   The unit of mass 1000 g = 1 kg

Granite   An igneous stone.

Gravity   The attraction of the earth for all solids, liquids and gasses.

Grout loss   The leakage of cement and water at formwork joints, bolt positions etc. Causes a surface defect having a sand textured appearance lacking in cement paste.

GRP   Glass reinforced plastic.

GS   *See* General structural.

Gypsum plaster   For internal use different grades of gypsum plaster are used according to the surface and coat. Undercoats use browning for general use and bonding for concrete; finishing coats, use finish on an undercoat or board finish for plasterboard.

HAC   *See* High alumina cement.

Hard water   Water which contains salts of calcium or magnesium. These insoluble salts tend to furr up in plumbing systems and cause restricted flow.

Harden   To become solid. May be applied to the curing or drying of adhesives, concrete, mortar and paint films etc.

Hardwood   Timber from broadleaf trees, which are mostly deciduous, such as oak, teak, beech and mahogany. Hardwoods have a fairly complex cellular structure consisting of pores, fibres and parenchyma. In addition they may also be classified as either ring or diffuse porous.

Heart   The centre of an object. *See also* Heartwood.

Heartwood   The inner, more mature part of a tree that no longer conveys the sap. Often darker in colour and more durable than the sapwood.

Heat   A form of energy that materials process. The higher their temperature the more energy they have. The energy is due to the movement of atoms and molecules, thus it is kinetic energy.

Heat capacity   The amount of heat, in joules, required to raise the temperature of a substance by one degree C or K. *See also* Specific heat capacity.

Hectare   An area of 10,000 m^2.

Hemihydrite   A gypsum plaster that has been heated to drive off some of its water and is therefore quick setting. *See also* Anhydrous.

High alumina cement   This uses bauxite (aluminium oxide) instead of clay. It develops very early strength which is much higher than OPC, although in the long term it has been found unstable due to conversion and is therefore now rarely favoured for structural work.

Homogeneous   A material having a uniformed composition throughout.

Honeycombing   A concrete surface defect resulting in a coarse stony surface with air voids, lacking in cement paste and fine aggregate. Can be caused through excessive leakage at formwork joints, a poor placing method, mix segregation or inadequate compaction. Also a timber defect.

Hooke's Law Law stating that strain produced in a member is proportional to the stress within the elastic limit. *See also* Modulus of elasticity.

Humidity A measure of the water vapour present in the air. *See also* Relative humidity.

Hydrate A compound that contains chemically combined water, generally in the form of water of crystallization.

Hydrated Containing chemically combined water, a hydrate the opposite of anhydrous. *See also* Hemihydrate and Hydrous.

Hydration The chemical reaction of a substance combining with water, as in the setting and curing of concrete, mortar and plaster.

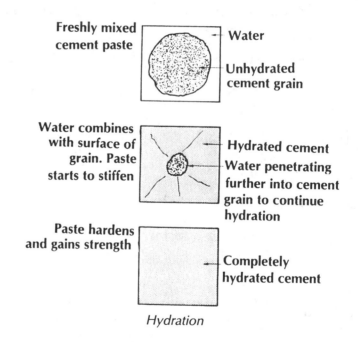

Hydration

Hydraulic cement A cement such as Portland cement which will set and cure under water.

Hydrophobic A water repellant. Hydrophobic cement is an OPC interground with a hydrophobic substance for use in damp, poor storage conditions to reduce the risk of bag setting.

Hydrostatic pressure The pressure exerted by a column of liquid due to atmospheric pressure e.g. in formwork due to fluid concrete and in plumbing systems due to heat.

Hydrous Containing water.

Hygrometer An instrument used to measure relative humidity.

Hygroscopic A solid which tends to absorb water vapour from the atmosphere and becomes wet, such as cement, plaster and timber. *See also* Bag set.

I

Ice Water in a solid state due to low temperatures, will start to form ice at 0°C. *See also* Melting.

Igneous stones These were formed by the original solidification of the earth's molten material (magma). They are extremely hard wearing and strong in compression. Granite is the main building example.

Ignition The lighting of a flammable vapour to commence combustion.

Ignition temperature The lowest temperature at which a solid material gives off a flammable vapour, at which stage it will start to burn providing a flame is present for ignition e.g. the ignition temperature of timber is about 300°C.

Immiscible Said of two liquids that do not mix together, but form into layers, one above the other, such as oil and water.

Impact sound *See* Sound insulation.

Impermeable A material that will not allow the passage of another substance through it e.g. a DPC, whereas a permeable one will. *See also* Porous and Impervious.

Impervious A material that will not allow the passage of a liquid through it but may be permeable to gases e.g. a breather paper.

Inert A substance which is not easily changed by a chemical reaction.

Insoluble A substance that will not dissolve in a solvent to form a solution, as opposed to a soluble one that will.

Insulation Any means of isolating heat, sound or electricity in a particular space. Also a material that is a good insulator used for such purposes. Also concerning fire resistance; the ability of an element to resist the passage of heat, which would increase the temperature of the unexposed face to an unacceptable level. *See also* Stability and Integrity.

Insulator A material that prevents or limits the transfer of heat and sound or electricity. Most non-metals and gases are good insulators. *See also* Thermal and Sound insulation.

Integrity Concerning fire resistance; the ability of an element to resist the passage of flames or hot gases from the exposed to the unexposed face. *See also* Stability and Insulation.

Interstitial condensation Internal condensation occurring within the structure of a building as opposed to surface condensation. Occurs when the temperature of the structure falls below the dew point.

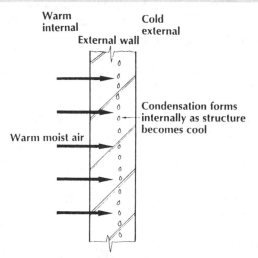

Interstitial condensation

Intumescent A material used in strips or in paints and varnishes. Used around doors for fire resistance and on other surfaces to reduce surface spread of flame. On heating up it expands and bubbles. This seals gaps, and also forms an insulating layer that cuts off a fire's oxygen supply.

Ion An atom, or group of chemically combined atoms, carrying an electric charge. Positive ions are called cations, negative ions are called anions.

J

Joule The unit of work or energy. *See also* Power.

K value   Thermal conductivity.

KAR   *See* Knot area ratio.

Kelvin   A unit of temperature. Its symbol is K.

   1 K = 1°C
   Absolute zero is 0 K which is −273°C

Kiln seasoning   The artificial seasoning of timber either in a compartment kiln, which is much like a large oven, or alternatively in a progressive kiln which is much like a heated railway tunnel.

Kilogramme (kg)   1000 grammes, the unit of mass.

Kilojoule (kJ)   1000 joules, the unit of work or energy.

Kilometre (km)   1000 metres, the unit of length.

Kilonewton (kN)   1000 newtons, the unit of force.

Kilowatt (kW)   1000 watts, the unit of power.

Kilowatt hour (kWh)   A practical unit of power, mainly used to measure the consumption of electricity. 1 kWh = 1 kW maintained for 1 hour.

Kinetic energy   The energy of a moving object.

Knot area ratio   The area of knots in the worst cross-section of a length of timber, compared with the cross-sectional size. Forms the basis of visual stress grading.

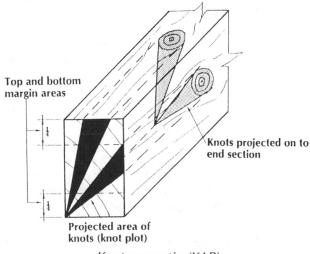

Knot area ratio (KAR)

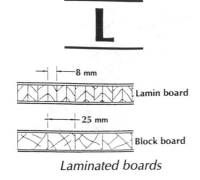

Laminated boards

Laminated boards A sheet material consisting of strips of timber which are glued together, sandwiched between two veneers. Used for panelling, doors and cabinet construction. Blockboard and laminboard are the two main types.

Laminboard *See* Laminated boards.

Latent heat The heat taken in or absorbed when a substance changes its state of matter. *See also* Latent heat of fusion and Vaporization.

Latent heat of fusion The heat absorbed by a solid as it turns into a liquid.

Latent heat of vaporization The heat absorbed by a liquid as it turns into a gas.

Lean concrete A mix having a high aggregate/cement ratio. *See also* Rich concrete.

Lever A simple machine that can turn about a pivot. It uses the principle of moments, normally to move a load using a smaller effort.

Light Electromagnetic waves which cause a sensation in the eye. Different wavelengths are seen as different colours.

Lightweight aggregate Having a density of not more than 1200 kg/m^3 for fine aggregates and not more than 1000 kg/m^3 for course aggregates.

Lime sand plaster Used for both under and finishing coats (rarely), although lime can be added to other plasters to improve their workability.

Liquefied The change in state of a substance into a liquid from either a gas or solid such as condensation and melting.

Liquefied gas A gas which has changed into a liquid state by lowering its temperature and/or pressurizing it. Used as a fuel or refrigerant e.g. butane and propane.

Liquid A state of matter midway between a solid and a gas. Liquids have molecules that are held by weaker forces than solids; thus liquids have a definite volume but no shape. A liquid will always take up the shape of its container.

Litre A unit of capacity or volume, 1000 litres is equal to one cubic metre.

Litmus A substance obtained from plants used as an acid base indicator. In contact with acids it turns red and with alkalis it turns blue. *See also* pH value.

Low soluble salts A clay brick classification (L) means that they are for use in exposed conditions.

Lubricant A substance which is used to reduce friction between moving solids e.g. oil.

Luminous Giving out light.

Lumen A measure of the amount of light received by a surface or emitted from a point source.

Lux A unit of illumination; one lumen per square metre.

Machine general structural (MGS) A machine stress graded timber, used for structural applications. General structural (GS) is a visually graded equivalent. *See also* Machine special structural.

Machine special structural (MSS) A machine stress graded timber used for highly stressed situations. Special structural (SS) is the visually graded equivalent. *See also* Machine general structural (MGS).

Mahogany A hardwood from West Africa and South America. Used for joinery, cabinet work, furniture, panelling, high-quality flooring and boat fitting. Its colour varies from pink to red/brown.

Malleable A metal that can be hammered into different shapes. *See also* Ductile.

Manometer An instrument for comparing pressures.

Manufactured aggregate Mostly created as a by-product of some industrial processes often involving heat e.g. blast furnace slag, pulverized fuel ash, exfoliated vermiculite and expanded perlite.

Mass A measure of how much material a substance has, or how heavy it is. Normally measured in kilogrammes (kg) or grammes (g). *See also* Density, Force and Weight.

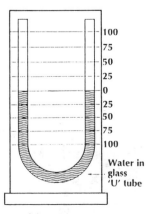

Manometer

Mastics Sealing compounds, either rubber, bitumen, or plastic based, which are used in modern buildings to seal joints around the outside of windows and doorframes, against rain, air, and sound, while still allowing differential movement between the two materials. These mastics are normally sold in tubes for use in guns which have a nozzle through which the mastic is extruded.

Mayonnaise paste A timber preservative applied as a paste in situation, to large, sectioned, structural timbers, which have been affected by fungi or wood boring insects. This paste slowly penetrates deep into the timber's heart making it far more effective than a liquid preservative, which in the situation would merely give a surface coating.

MDF *See* Medium density fibreboard.

Mechanical adhesion The penetration of an adhesive into the surface of the materials to be bonded. *See also* Adhesion.

Mechanical advantage In levers and other simple machines the difference between the effort put in and the load moved. Normally a load bigger than the effort applied is able to be moved.

$$\text{Mechanical advantage} = \frac{\text{load}}{\text{effort}}$$

Mechanics The study of forces and their effects. Statics are in equilibrium, dynamics are not.

Medium density fibreboard (MDF) A fibreboard, not to be confused with medium fibreboard. It is made by a dry process with synthetic resin binders and is used for fitments etc.

Medullary rays A group of parenchyma cells.

Melamine formaldehyde A thermosetting plastic used for adhesives and electrical fittings.

Melting The change of state from a solid to a liquid on heating. Water melts at 0°C which is the same as the freezing point of water.

Meniscus The curved surface of a liquid in a pipe, or small container, caused by adhesion e.g. water in a glass jar will rise at the edge as the adhesion of water to glass is stronger than the cohesion of water.

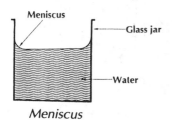

Meniscus

Metals   Metals are minerals; very few are found as such in nature. Most have to be extracted from metallic ores by either smelting in a furnace and/or eletrolytic methods. All metals are classified as being either ferrous or non-ferrous metals.

Metamorphic stone   These were formed from older stones that had been subjected to very high temperatures and pressures causing a structural change to take place. The two main examples are marble which was formed from limestone and slate which was formed from clay.

Metre   The unit of length. The symbol m.

MGS   *See* Machine general structural.

Mixture   A material created by mixing two or more substances together with no chemical change. They can be in any proportion, and can be separated easily by physical means. Each substance also retains its own properties. *See also* Compound.

Moderately frost resistant   A clay brick classification (M) means that they are durable except when saturated and subjected to repeated freezing and thawing.

Modulus of elasticity   The ratio of stress to strain which remains constant up to the elastic limit. Also termed Young's Modulus when using tension and compression. *See also* Hooke's Law.

Moisture   Water or other liquid in small quantities, often as a vapour. *See also* Moisture vapour.

Moisture barrier   *See* heading under Building construction.

Moisture content   The amount of water contained in the voids of a porous material, usually expressed as a percentage of its dry weight. May be determined by calculation after drying a sample using:

$$\text{Moisture content} \% = \frac{\text{wet mass} - \text{dry mass}}{\text{dry mass}} \times 100$$

or alternatively by use of a moisture meter.

Moisture meter   A device used to determine the moisture content of materials, by measuring the electrical resistance between two points of an electrode which are inserted into the material.

Moisture vapour Water that is in a gaseous state of matter due to either evaporation or boiling. *See also* Condensation and Humidity.

Molecule A group of two or more atoms joined together. The smallest part of a compound that can exist independently while still retaining the compound's properties.

Molten A solid in a liquid state normally as a result of heating.

Moment The turning effect of a force about a fulcrum. It is the product of the force and its distance from the fulcrum.

Moment = force × distance

Moments can be considered to have either a clockwise or anti-clockwise turning effect. Where moments are balanced or equal they are said to be in a state of equilibrium. *See also* Principles of moments.

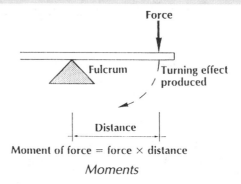

Moments

Monomer Small molecules that can join to form a larger molecule called polymer.

Mortar The gap-filling adhesive that holds bricks, blocks or stonework together to form a wall. It takes up the slight difference in shape and provides a uniform bed to transfer the loads from one component to the next.

Mould Any small fungi which feed on organic matter, often found in conjunction with dampness and condensation. *See also* heading under Building construction.

MSS *See* Machine special structural.

Natural aggregate   Naturally occurring materials e.g. gravel, sand and crushed rock.

Natural seasoning   The natural drying of timber in the air. Involves stacking timber in open-sided, roofed sheds to protect it from the sun and rain but allow a free circulation of air. A moisture content of 18 to 20 per cent can be achieved in two to twelve months. *See also* Artificial seasoning.

Neutral   A solution which is neither acidic or alkaline, having a pH value of seven. Also a particle having no electrical charge e.g. neutron. *See also* Litmus. (268)

Neutralization   The process in which either the pH of an alkaline solution is decreased to seven or the pH of an acidic solution is increased to seven. The resulting solution is neutral. *See also* Litmus.

Neutron   A neutral particle that exists in the nucleus of many atoms.

Newton   The unit of measurement for a force. It is the force which will accelerate a mass of 1 kg by 1 m/s².

Non-combustible   A solid material that is incapable of burning, as opposed to a combustible one which does burn. *See also* Non-flammable.

Non-ferrous   A metal containing no iron, such as zinc, copper, tin, lead, aluminium and magnesium. They corrode slowly when exposed to sulphur-containing gases or solutions which are present in the air. This corrosion forms a protective film which prevents further corrosion. Most principal non-ferrous metals are rather soft; they are often mixed in their molten state to produce harder alloys. *See also* Ferrous.

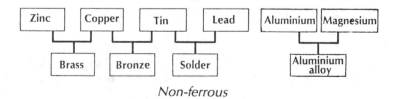

Non-ferrous

Non-flammable   A liquid or gas that is incapable of burning as opposed to a flammable one which does burn. *See also* Non-combustible.

Normal soluble salts   A clay brick classification (N) means that they are liable to cause efflorescence (a white powdery stain on the face of brickwork caused by the surfacing of a soluble mineral salts).

Not frost resistant   A clay brick classification (O) means that they are recommended for internal use only.

Nucleus   The central part of an atom.

Nylon   A thermoplastic used for door and window fittings.

Oak   A hardwood from Europe, America and Japan used for panelling, external joinery, fence posts, internal fitments, church fittings and, traditionally, large section floor and roof timbers. It is light brown in colour with silver flecks or streaking.

Ohm   The unit of electrical resistance. A resistance of 1 ohm exists between two points in a circuit when a potential difference of 1 volt between the points produces a current of 1 ampere. *See also* Ohm's Law.

Ohm's Law   The electric current flowing through a conductor is proportional to the voltage between its ends. Resistance is determined from this. $R = V/I$ where R is resistance in ohms; V is in volts; and I in amperes.

Opaque   Any solid or liquid through which light cannot travel. *See also* Translucent and Transparent.

OPC   *See* Ordinary Portland cement.

Organic   Containing carbon compounds, organic substances are normally formed by animals and plants or are residues of these organisms. *See also* Mould and Fungi.

Organic solvent   A timber preservative consisting of toxic chemicals which are mixed with a spirit that evaporates after application. Generally considered to be superior to tar oils and water-soluble preservatives.

Oxidation   The combining of oxygen with a material such as combustion and the rusting of ferrous metals. *See also* Oxide.

Oxide   A compound consisting of one element combined with oxygen. Rust is an oxide of iron.

Paint A thin decorative and/or protective coating which is applied in a liquid or plastic form and later dries out or hardens to a solid film, covering a surface. Paints consist of a film former, known as the vehicle; a thinner or solvent (water, white spirit or methylated spirit etc.) to make the coating liquid enough; and a pigment suspended in the vehicle to provide covering power and colour. Paint schemes require either the application by brush, spray or roller, of one or more coats of the same material (varnish, emulsion and solvent paints) or a build up of different successive coats, each having their own functions (primer, undercoat and finishing coat).

Parenchyma A timber cell present in both hardwoods and softwoods. Also known as ray parenchyma or medullary rays. These are food storage cells that radiate from the centre of the tree trunk like the spokes of a wheel. *See also* Tracheids, Pores and Fibre. (269)

Particle A molecule or an atom, any small piece of material such as a grain of sand, also any substance with a small volume.

Particle boards A sheet material mainly known as chipboard, manufactured from woodchips and flakes impregnated with an adhesive. Used for flooring, furniture, and cabinet construction.

Pascal A unit of pressure (symbol Pa) equal to 1 N/m^2 (newton per square metre) 100,000 Pa = 1 bar.

Paver A brick used for paving.

Perforated brick A brick with holes passing through it, up to 25 per cent of its gross volume.

Permeable A material that allows another substance to pass through as opposed to an impermeable one that will not. *See also* Porous and Impervious.

Permissible stress The stress that can safely be carried by a structural timber member after making an allowance to the grade stress for duration of load and moisture content.

pH value A value on a numerical scale ranging from 0 to 14 indicating how acidic or alkaline a solution is. At 7 the solution is neutral. Between 7 and 0 the solution's acidity increases, between 7 and 14 the solution's alkalinity increases. *See also* Litmus.

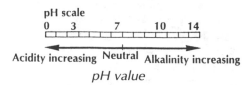

pH value

Phenol formaldehyde   A thermosetting plastic used for adhesives and electrical fittings.

Photosynthesis   A process which occurs in the leaves of a tree, where the green substance (chlorophyll) absorbs sunlight and carbon dioxide and gives off oxygen. As a result a sugar is formed which is added to the water and minerals which have travelled up from the roots.

Physical change   A change in the state of a material such as ice melting, or a change in the physical properties of a material such as the expansion of metals on heating. Neither alters chemical properties.

Plain concrete   A concrete mix without any admixtures or reinforcement.

Plaster   The material applied on internal walls and ceilings to provide a jointless, smooth surface which can be easily decorated. External plastering is normally called rendering. The plaster is a mixture that hardens after application; it is based on a binder (gypsum cement or lime) and water, with or without the addition of aggregates. Depending on the background (surface being plastered) plastering schemes may require the application of either one coat, or undercoats to build up a level surface followed by a finishing coat.

Plasterboard   A sheet material comprising a gypsum plaster core sandwiched between sheets of heavy paper, used for wall and ceiling linings.

Plastic   Plastics are synthetic (man-made) materials derived mainly from petroleum and also coals. Plastic products are formed into their required shape while the solution is still in its soft state, by a variety of processes including extrusion through a shaped die, film blowing into a tube, sheet casting onto a cooled surface and injection or vacuum forming into a mould. Two distinct groups of plastics can be identified: thermoplastics and thermosetts.

Plastic laminate   A sheet material made from layers of paper impregnated with an adhesive, used for worktops and other horizontal and vertical surfaces requiring decorative, hygienic and hard-wearing properties.

Plasticizer   A solution added to the mixing water of concrete, plaster and mortar mixes to lower the surface tension. This improves dispersion of the cement particles and lubricates the cement paste.

Plywood   A sheet material, usually consisting of an odd number of thin layers glued together with their grains alternating. Used for flooring, formwork, panelling, sheathing and cabinet construction.

3 ply (equal thickness layers)

Stout-heart (thicker core)

Multiply (over 3 layers)

Plywood

Polyester   A thermosetting plastic reinforced with glassfibre to form glass reinforced plastic (GRP).

Polymer   A large molecule made by joining together smaller molecules or monomers.

Polystyrene   An air-expanded plastic resin used in sheets or loose fill for wall and roof insulation.

Polythene   A thermoplastic material used for damp proof courses and membranes, waste pipes, water pipes and tanks.

Polyurethane   A thermosetting plastic used for foam products and mastic sealants.

Polyvinyl chloride (PVC)   A thermoplastic used for electrical insulation, sheets and floor tiles. Unplasticized or rigid PVC known as UPVC is used for stackpipes, wastepipes and gutters.

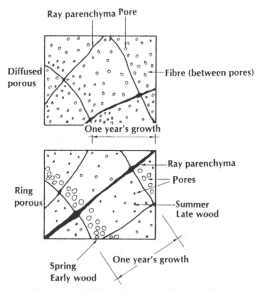

Pore (magnified hardwood section)

Pores The cells of hardwood that conduct the rising sap, also termed vessels. They may be evenly distributed throughout a year's growth where they are termed diffuse porous hardwood; or they may form a definite ring of large cells in the spring growth, with smaller cells spread through the remainder of the growing season. The latter are termed ring porous hardwoods. *See also* Fibre and Parenchyma.

Porous A material that is full of small air pockets through which liquids may pass, hence it is permeable.

Potential difference The difference in the electric force between any two points in an electric circuit. Normally termed voltage.

Power The rate of doing work. The amount of work done divided by the time taken. If a force does 500 J of work in 5 seconds, then the power is 500 J divided by 5 s = 100 J/s or 100 W since the transfer of one joule per second is equal to a watt. One horse power is approximately 750 watts. *See also* heading under General.

Precipitate A solid that forms within a solution.

Preservatives Liquids applied to timber in order to poison the food supply of fungi and wood-boring insects. The main types are: tar oils, water-soluble, organic solvent and mayonnaise paste.

Pressure The force that is applied over a particular area. Force is measured in newtons and area in square metres. So pressure can be expressed in newtons per square metre (N/m^2):

Pressure = force/area

Consider a force of 150 N acting on an area of 2 m^2. The pressure will be 75 N/m^2. This pressure may also be expressed in pascals as 75 Pa since Pa = 1 N/m^2.

Pressure impregnation The application of a timber preservative under pressure, which achieves almost full penetration of the cells. More effective than non-pressure treatments such as brushing or spraying which merely form a surface coating.

Pre-stressed concrete A structural concrete unit that has been given a high tensile strength by embedding tensioned wires or cables into it. In pre-tensioned the wires are stretched before the concrete is cast whereas in post-tensioned the wires are tensioned after the concrete has hardened.

Primary cell A device that produces an electric current, by means of a chemical reaction. A simple cell consists of two different metals coupled together in an electrolyte.

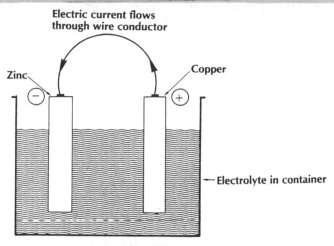

Electric current flows
through wire conductor

Zinc

Copper

Electrolyte in container

Primary cell

Primer The first coat of paint on a surface. May form a protective coat against moisture and corrosion or act as a barrier between dissimilar materials. Also provides a good surface for subsequent coats.

Principle of moments A statement used in structural design to determine support reactions. For a body to exist in a state of equilibrium the sum of the clockwise moments (CWM) must equal the sum of the anti-clockwise moments (ACWM).

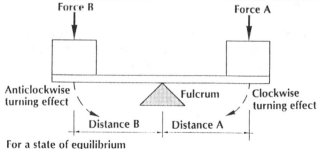

Force B

Force A

Anticlockwise
turning effect

Fulcrum

Clockwise
turning effect

Distance B

Distance A

For a state of equilibrium

Clockwise moments (CWM) = anticlockwise moments (ACWM)

Force A × distance A = force B × distance B

Principle of moments

PVC *See* Polyvinyl chloride.

Q

Quarrel or quarry *See* heading under Architectural style.

Quartering The reduction of a material sample into four equal parts, two of which are remixed in order to obtain a representative sample. Also the conversion of a tree into four sectors.

R

R value Thermal resistance.

Radial cut A piece of timber cut so that its annual rings meet the wider surface of the piece, throughout its width at an angle of forty-five degrees or more as opposed to tangential cut, where the annual rings are less than forty-five degrees. Used to produce best quality timber for joinery purposes. There is very little tendency to shrink or distort.

Radiation The transfer of heat from a hot object through space to a cold object. Termed radiant heat. Unlike conduction and convection it does not require a material or air for the transfer.

Ray *See* Parenchyma.

Reaction In mechanics the opposite effect to an action e.g. the force that supports a force. *See also* Equilibrium and Chemical reaction.

Ready-mixed concrete Concrete that is batched and mixed by a ready-mixed specialist and transported to a site in mixer trucks, as opposed to mixing on site.

Reinforced concrete A plain concrete that has had steel reinforcement embedded into it to increase its tensile and strength.

Reinforcement Steel embedded into concrete to resist tensile and shear stresses.

Relative density The relation of a material's density to the density of water. For concrete, density is 2400 kg/m^3, for water 1000 kg/^{3}, thus relative density = 2400 ÷ 1000 = 2.4 (there are no units for relative density). Thus concrete is 2.4 times heavier than water. Materials with a relative density of less than one will float in water. Those over one will sink. *See also* Buoyancy.

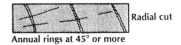

Annual rings at 45° or more

Annual rings at less than 45°

Radial cut

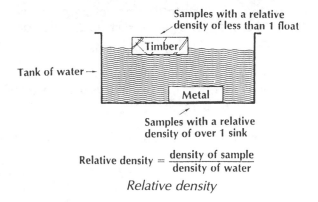

Relative density

Relative humidity   A measure of the air's humidity or moisture content compared with saturated air at the same temperature. Can be measured with a hygrometer. Saturated air had a relative humidity of 100 per cent. This is also known as the dew point. Air that is half saturated has a relative humidity of 50 per cent. If the temperature is raised the relative humidity will go down since warm air has a greater capacity for absorbing and holding moisture than cold air. Condensation occurs when air with a high relative humidity is cooled.

Release agent   A substance that is applied to formwork in order to prevent adhesion between a concrete surface and the form face.

Resistance   The opposite of conductivity. That which opposes the flow of electricity through a conductor, measured in ohms. A resistance of one ohm exists between two points in a circuit when one volt produces a current of one ampere. Resistance = volts/amperes. *See also* Ohm's Law.

Resorcinol formaldehyde   A thermosetting plastic used for adhesives and electrical fittings.

Retarder   A substance that slows down a chemical reaction. Sometimes added to adhesives, concrete and mortar to slow the rate of curing or hydration.

Reverberation   The repeated reflection of sound from hard surfaces, such as an echo.

RHPC   *See* Rapid hardening Portland cement.

Rich concrete   A concrete mix having a low aggregate/cement ratio. *See also* Lean concrete.

Ring porous   *See* Pores.

Rust   An oxide of iron. *See also* Corrosion.

S

Sacrificial coating A surface coating, often zinc, which is applied to iron and steel to give it sacrificial protection. *See also* Galvanized.

Sacrificial protection The coupling of two metals so that one is sacrificed by electrolytic corrosion in order to protect the other. *See also* Galvanizing.

Salt A compound formed when an acid reacts with a base.

Sapwood The newly formed outer growth layers of a tree which conduct the rising, unenriched sap up to the leaves where photosynthesis occurs. *See also* Heartwood.

Saturated Unable to take any more. Air that is saturated is at the dew point. It has 100 per cent relative humidity.

Saturated solution A solution that has dissolved as much of a solid as is possible.

Scots Pine European redwood.

Seasoning The controlled drying of converted timber by natural (air seasoning) or artificial (kiln seasoning) means in order to achieve a moisture content which is in equilibrium with the surroundings in which the timber will ultimately be used.

Second seasoning The traditional practice of loosely framing up high-quality joinery, stacking and storing at an equilibrium moisture content, for it to condition. Should any defects occur during second seasoning the component can easily be replaced, prior to final gluing up.

Sedimentary stones These were formed by a sedimentary or settling process, whereby rock and organic particles were deposited, often on the seabed, in layers which subsequently compacted to form a solid. Sandstone and limestone are the main examples. Being sedimentary they can contain hard and soft layers which weather at different rates.

Shaking A timber defect taking the form of splits along the grain of a piece of timber, particularly at its ends, it is the result of drying out too fast during seasoning.

Setting The change of state from a liquid to a solid. Often refers to the initial stiffening process. *See also* Curing.

Shear A stress that occurs when one part of a structural member tends to slip over another part. It has a slicing effect. Occurs midway between compression and tension and also near supports.

Sherardizing The coating of iron or steel with sacrificial coating. The iron or steel is heated with a zinc dust to a temperature slightly below the zinc's melting point, forming zinc–iron alloy internal layers with a pure zinc coating. *See also* Galvanized.

Sieve analysis A test to determine the grading of an aggregate.

Simple cell *See* Primary cell.

Simultaneous ignition The ignition of many flammable vapours at the same time. Occurs during the development of a building fire or combustion. Termed flashover.

Slump test The most common on-site test for workability. Involves measuring the distance that a standard cone of concrete collapses when its support is removed. Useful for comparing the consistency between different mixes.

Fill cone in three layers, tamping each thoroughly

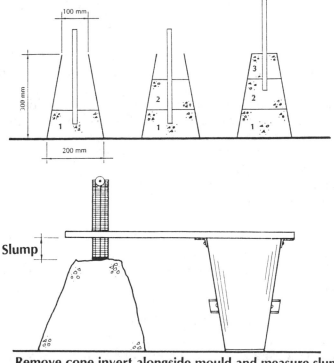

Remove cone invert alongside mould and measure slump

Slump test

Snots Hardened grout runs/fins hanging from the underside of soffits. The removal of snots prior to soffit finishing is termed 'snotting'; caused through the use of damaged or poorly jointed decking.

Softwood Timber from coniferous trees which are mostly evergreen such as European redwood, Douglas Fir and Spruce. Softwoods have a simple cellular structure consisting of tracheids and parenchyma.

Solid A state of matter in which molecules are held by strong forces. They do not have the same freedom of movement that liquids and gases have. Solids have a definite shape and a definite volume.

Soluble Able to dissolve in a solvent to form a solution, as opposed to an insoluble substance that cannot.

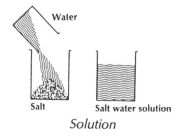

Water

Salt

Salt water solution

Solution

Solution A mixture normally obtained by dissolving a solid in a liquid. *See also* Soluble, Solvent and Insoluble.

Solvent A liquid in which a solid can be dissolved to form a solution.

Solvent paint Based on rubber, bitumen or coal tar and used for protecting metals and waterproofing concrete etc.

Sound insulation The design of a building to restrict the passage of airbourne and impact sound. Airbourne sound is vibrations in the air and requires either a solid mass construction to reflect sound waves, or the inclusion of lightweight porous materials that reflect sound waves to and fro between their pores, thus reducing their strength. Impact sound is vibrations that are connected directly to the structure such as from machinery or footsteps, it requires a discontinuous construction. Soft floor finishes can help to reduce the sound of footsteps.

Special structural (SS) A visually stress graded timber used for highly stressed situations. Machine special structural (MSS) is the machine graded equivalent. *See also* General structural (GS).

Specific gravity *See* Relative density.

Specific heat capacity The amount of heat in joules required to raise the temperature of one kilogramme of a substance by one degree. *See also* Heat capacity.

Spectrum   The result of passing white light through a prism; it splits into colours – red, orange, yellow, green, blue, indigo and violet.

Spontaneous combustion   Combustion that starts to burn without an ignition source; requires temperatures far higher than the material's ignition temperature or flashpoint.

Spore   A seed from which, given the right conditions, a fungus will start to grow.

Springing   A timber distortion being a curvature along the edge of a piece where the face remains flat.

Springing

Spruce   A softwood from both Europe and Canada/USA. Also known as whitewood. Used for general carpentry and joinery work, although it is not durable for use in external joinery and cladding. Has a pale white colour with little difference between sapwood and heartwood.

SRPC   *See* Sulphate resisting Portland cement.

SS   *See* Special structural.

Stability   Concerning fire resistance; the ability of an element to resist collapse. *See also* Integrity and Insulation.

State of matter   One of three physical states in which matter can exist. All materials are either solids, liquids or gases.

Static electricity   Electricity at rest as opposed to current or moving electricity. Caused by friction when items are rubbed together.

Statics   The study of objects under the action of forces but in equilibrium rather than dynamics.

Sterilize   To free a surface from all mould or fungi, normally by the application of a sterilizing solution.

Sterilizing solution   A toxic solution used to sterilize a surface. *See also* Preservative.

Stone   Natural stones or rocks are used extensively in building for walling, roofing, paving, and tiling. They may be classified into three groups: igneous, sedimentary and metamorphic.

Stoneware *See* Ceramic.

Strain The change in the shape or size of a member when stressed.

Stress A body subjected to a force is in a state of stress. Stress is the result of force per unit area acting on a solid causing strain:

$$\text{Stress} = \frac{\text{force}}{\text{area}}$$

The units of stress are newtons per square metre (N/m^2). The stress that breaks a structural member is termed its ultimate stress; the stress below which only a small percentage of samples will fail is termed the characteristic stress; the working or basic stress is the stress that can be safely sustained after making an allowance for accidental overload in use (factor of safety). *See also* Grade and Permissible stress, Compression, Tension and Shear.

Stress grading The strength grading of timber for structural purposes. Can be carried out visually where it is based on the amount and location of defects, or by machine where it is based on deflection under load. *See also* Knot area ratio, General structural and Special structural.

Substance Any chemical or mixture of chemicals.

Suction The raising or drawing in of a fluid to produce a partial vacuum which is filled by atmospheric pressure.

Sulphate resisting Portland cement A Portland cement for use underground in high sulphate conditions.

Surface condensation Moisture forming when warm moist air meets a cool surface. Occurs due to the air temperature falling below its dew point.

Surface tension The tendency of a liquid to act as though it has a skin covering, holding it together. This is due to the cohesive forces on the molecules pulling towards the centre. *See also* Capillary attraction.

Surfactant A substance added to a liquid in order to increase its spreading and wetting power.

Suspension A mixture consisting of small pieces of a solid distributed in a liquid.

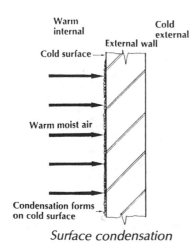

Surface condensation

Tangential cut A piece of timber cut so that its annual rings meet the wider surface of the piece, over at least half its width, at an angle of less than forty-five degrees. As opposed to radial cut, where the annual rings are at forty-five degrees or more. Used to produce structural timber, but is prone to distortion.

Tarnish The corrosion or surface discolouration of bright non-ferrous metals, termed verdigris on copper.

Teak A hardwood from India, Burma and Thailand. Suitable for most high-quality joinery uses. Specially suitable for use near chemicals as it is extremely resistant. It is pinkish/brown in colour, often with greenish tints.

Temperature A measure of how hot or cold something is. Normally using a thermometer graduated in units of degrees celsius (C) or kelvin (K).

Tension Stress in a structural member that tends to pull or stretch. It has a lengthening effect. The opposite of compression. *See also* Shear.

Terracotta *See* Ceramic.

Thermal conductivity The K value of a material, it is a measure of a material's ability to conduct heat. May be expressed as the flow of heat, in watts, through a square metre of material, one metre thick, with a temperature difference of one degree C or K between its inside and outside surfaces. A material with a high K value is a good conductor and a poor insulator, such as steel with a K value of 48. Mineral wool with a K value of 0.04 is a good insulator but a poor conductor. *See also* Thermal resistance and Thermal transmittance. Units for K value are W/M°C or K.

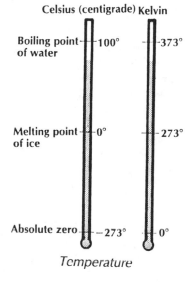

Temperature

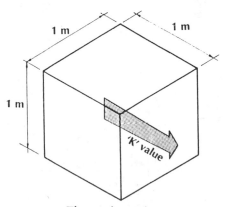

Thermal conductivity

Thermal expansion   The increase in length, area or volume of a material caused by a rise in temperature. *See also* Coefficient of thermal expansion.

Thermal resistance   The resistance, known as R value, of a material to the flow of heat per square metre for any given thickness. It can be determined by dividing a material's thickness in metres by its K value (thermal conductivity):

$$R \text{ value} = \frac{\text{Material thickness in metres}}{\text{K value}}$$

Units for R value are $M^2 \, {}^{\circ}C$ or K/W. *See also* Thermal transmittance.

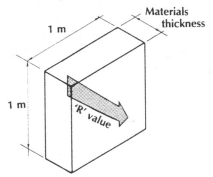

Thermal resistance

Thermal transmittance   The rate of heat flow through an element of a building one metre square, known as the U value. The U value is the reciprocal of the sum of R values of the materials in the element, plus standard resistances for the air clinging to the inside and outside surfaces and any cavities.

$$U \text{ value} = \frac{1}{\text{sum of resistance (R)}}$$

Units for the U value are $W/M^2 \, {}^{\circ}C$ or K. *See also* Thermal conductivity.

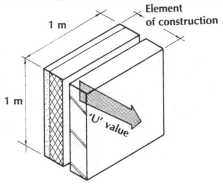

Thermal transmittance

Thermometer   A device used to measure temperature.

Thermoplastic   A plastic that softens on heating and can be moulded. This process can be repeated on subsequent heating. Also applied to synthetic resin adhesives which soften on heating, or following the addition of a solvent. *See also* Thermosetting.

Thermosetting   A plastic that undergoes a chemical change when setting. It cannot be softened again. Also applied to synthetic resin adhesives which, once set, will not soften on heating or following the addition of its solvent. *See also* Thermoplastic.

Thixotropic   The property of a liquid that can alter its viscosity, such as a paint that thickens in its can after storage but thins again after stirring.

Toxic   A substance that is poisonous to animal or plant life, used in preservatives and sterilizing solutions.

Tracheid   A box-like timber cell which forms the main structural tissue of softwood. Tracheids provide the timber with its mechanical strength and also conduct the rising sap. *See also* Parenchyma.

Translucent   Any solid or liquid through which light can travel. *See also* Transparent.

Transparent   Any solid or liquid through which light can travel and which allows an observer to clearly see objects on the other side. *See also* Translucent.

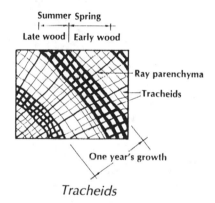

Tracheids

U value   Thermal transmittance.

Undercoat   A paint used on a primed surface to give it a uniform body and colour on which a finishing coat can be successfully applied. Also a plaster coat applied under the finish.

Unplasticized polyvinyl chloride   *See* Polyvinyl chloride.

Unsaturated   A solution capable of dissolving further solute or a gas capable of containing more liquid. *See also* Soluble and Solvent.

UPVC   *See* Polyvinyl chloride.

Urea formaldehyde A thermosetting plastic used for adhesive and electrical fittings.

Vacuum A space with nothing in it. A partial vacuum is achieved when as much air, or other gases, as possible have been removed from a space.

Vapour A gas which is normally a solid or liquid. *See also* Boiling point, Condensation and Evaporation.

Varnish A paint without a pigment, used for clear finishing.

VB consistiometer test A concrete workability test more suited to a laboratory than on site. It measures the time in seconds (V-B degrees) taken to compact a standard cone full of concrete.

Veneer A thin piece of timber peeled or cut from a log, used in plywoods and for facing other sheet materials.

Ventilation A free circulation of air or the introduction of air into a space. May be achieved by natural means, such as an air brick, window or door, or by mechanical means such as fans. *See also* Air conditioning.

Verdigris The green corrosion that forms a protective coat on copper. Also termed tarnish.

Vermiculite A flaky mineral which is expanded or exfoliated at high temperature, used as a loose fill insulation for roofs. Also added as a lightweight aggregate to concrete and plaster, improving their thermal insulation and fire-resistant properties.

Vessels *See* Pores.

Viscosity The thickness of a liquid which determines its flowing properties. High viscosity or thick liquids resist flow whereas low viscosity or thin liquids flow readily. *See also* Thixotropic.

Vitreous china *See* Ceramic.

Voids Air spaces in concrete formed between aggregate particles.

Volt   The unit of electric potential. One volt is the difference in potential energy between two points, if one joule of work is done, transferring one coulomb of charge.

Voltage   A potential difference measured in volts.

Volume   The amount of space that a substance occupies. Measured in cubic metres (m^3). *See also* Capacity and Litre.

Water/cement ratio   The relationship between the amounts of water and cement in a concrete mix.

Water of crystalization   The water that is chemically bonded within crystals. *See also* Hydration.

Water soluble   A substance that will dissolve in water.

Water vapour   *See* Moisture vapour.

Watt (W)   The unit of power. One watt is produced when one joule of work is done in one second i.e. one watt = one joule divided by one second. Also the energy of an electric current. Electric appliances are rated in watts by their consumption of energy.

   watts = amperes × volts

For example for a ten amp current at 240 W = 10 × 240 = 2400. *See also* Kilowatt.

Wavelength   The distance between two corresponding points in a wave, such as the distance between vibrations in the sound wave.

Weight   The force of gravity on a mass which attracts it towards the earth. As weight is a force it is measured in newtons.

Wet rot   A fungi that feeds on, and destroys, really wet timber rather than just damp timber. Most often found in cellars, neglected external joinery and rafter ends. Causes timber to soften, darken, develop cracks along the grain and lose strength. *See also* Dry rot.

White or coloured Portland cement   A white or coloured Portland cement made using white china clay; pigments are added for coloured cements.

Whitewood   Spruce.

Winding   A timber distortion where the member is twisted. Also framed joinery is said to wind where it is twisted.

Winding

Wood-boring insects   Five main species of insects, the larva of which bore or eat into the timber, for sometimes many years, causing structural damage e.g. furniture beetle, deathwatch beetle, lyctus beetle, longhorn beetle and weevils.

Woodwool slab   A sheet material made from shavings coated with a cement slurry (roof decks, permanent formwork).

Woodworm   The larva of wood-boring insects, which bores into the wood causing structural damage.

Work   A name for energy transfer. Work is done when a force moves an object. The amount is measured in joules (J) and is obtained from: work done = strength of force × distance moved. For example if 100 N is moved 5 m work done = 100 × 5 = 500 J. *See also* Watt.

Young's Modulus   *See* Modulus of elasticity.

6
SERVICES
AND FINISHES

Access eye   A small sealable opening in a pipe, often at bends or traps, to permit cleaning or rodding out. Also known as cleaning eye or rodding eye.

After flush   The small quantity of water that slowly trickles down into a WC pan to remake the seal, after the main flush.

Air change   The number of times that the volume of air in a room is changed in order to keep it adequately ventilated. This will depend on the type and use of room e.g. a bathroom will require more frequent air change than a lounge. Normally specified as air changes per hour.

Air conditioning   The production and maintenance of a desirable internal atmospheric environment, irrespective of any external conditions. Air conditioning is more than merely mechanical ventilation as, in addition to providing air changes, it also has the ability to raise or lower the temperature and control humidity.

Air lock   A stoppage in the flow of water in pipework due to a trapped air bubble. Mainly occurs when refilling previously drained hot water systems containing a hot water storage cylinder or in radiators. *See also* Air vent and Bleeding.

Air test   The testing of a length of drain for water tightness using air pressure. *See also* Manometer and Water test.

Air vent   A device located at the highest point in pipe runs and radiators for releasing trapped air. *See also* Radiator key.

Ambient   The surrounding conditions such as temperature.

Anneal   To soften metal by applying heat. *See also* Annealing under Materials and scientific principles.

Anti-siphon pipe   An airpipe in a drainage system to prevent induced siphonage. This is connected at one end to the ventilation pipe and at the other, the downstream side of traps, to sanitary appliances.

Anti-siphon trap   A deep seal trap which is almost impossible to draw by induced siphonage because of the large amount of water it contains.

Apron flashing   A one-piece flashing used to weather the lower side of a chimney stack where it penetrates a pitched roof. *See also* Back gutter.

Automatic flushing cistern   A cistern that is designed to flush at predetermined intervals. Mainly used to cleanse urinals.

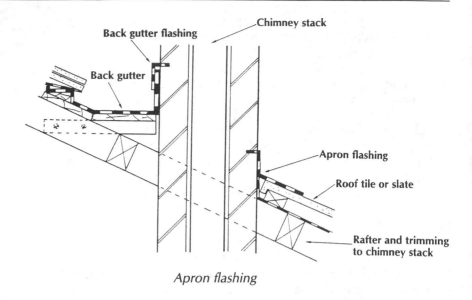

Apron flashing

B

Back boiler   A small boiler fitted in the recess behind either an open fire, room heater or gas fire. Provides domestic hot water and in some circumstances may also heat radiators for central heating.

Back drop   A connection to an inspection chamber in which the branch is at a higher level than the main drain and enters via a vertical pipe. This avoids excessive excavation or fall.

Back flow   Water flow in the opposite direction to that intended.

Back gutter   A gutter or flashing between the upper edge of a chimney stack and an abutting sloping roof. *See also* Apron flashing and Step flashing.

Back-inlet gulley   A trapped drainage gulley, with either a horizontal or vertical inlet, arranged to receive a rainwater pipe, or a wastepipe above the level of the water seal but below the sealed cover or grating.

Background   The surface on which the first coat of plaster or paint is to be applied.

Balanced flue   An arrangement whereby the air intake and the combustion flue outlet are combined in one outside terminal. A method commonly used for domestic gas-fired room sealed appliances.

Balancing valve   A lockshield valve.

Ball cock   A ball float valve.

Ball float valve   A valve used in water cisterns, operated by a lever arm and float which shuts off the water supply when the float reaches a predetermined level.

Banker   A board on which plastering materials are mixed.

Bayonet holder   The part of a light fitting that receives the end of the lightbulb. *See also* Ceiling rose.

Bead   A metal trim used in plastering and rendering to reinforce corners or stopped ends.

Bedding   The fixing of fibrous plaster mouldings to their background using an adhesive.

Belfast sink   A traditional deep-sided stoneware sink.

Benching   The sloping surfaces formed on either side of an inspection chamber channel. Their purpose is to ensure that solids flow/roll back into the channel and are not left behind after a blockage or flood.

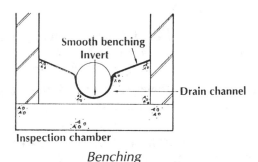

Benching

Bend   A curved length of pipe, tube, conduit or channel.

Bib tap   A tap with a horizontal inlet connection rather than the normal vertical type. Used over Belfast sinks where the taps project from the wall face.

Bidet   A sanitary appliance used for washing the excretory organs. May also be used as a footbath.

Bittiness   A defective paint finish showing bits of dust etc. in the paint film.

Bleeding   A paint film defect which allows the surface beneath to show through; the draining of water from a plastered surface before hardening; also the venting of radiators to remove trapped air.

Blistering   A paint film or plastering defect showing as hollow projections.

Bloom   A paint defect resulting in a dull film on gloss-painted finishes.

Boiler   A water heater.

Bonded   The practice of connecting water and gas service pipes to the main electrical earthing terminal/electrode.

Bossing   The process of shaping malleable materials e.g. aluminium and lead.

Bottle trap   A wastepipe trap in the shape of a bottle, rather than the normal U trap. It has an inlet below the water line of the trap and a side outlet.

Branch   A subsidiary drain to the main drainage run.

Branch fitting   A Y-shaped drainage channel which connects a branch to the main drain in an inspection chamber.

Brushing out   The initial application of paint with a brush in order to achieve a maximum even coverage before laying off.

Burning off   The removal of old paint and varnish films by softening with a heat source and subsequently scraping off.

Bus bar   A bare electrical conductor carrying a heavy current which is used to distribute power to other circuits. Found in consumer units and distribution boards.

Bush   A pipe fitting used to join together pipes of different diameter.

Butt jointing   The most common method of hanging wall coverings, their edges just touching without a gap or overlap.

Bypass An arrangement of pipes for water or gas which directs the flow around, instead of through, a particular pipe or piece of equipment.

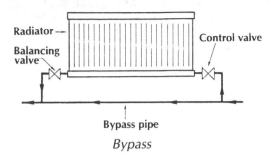

Bypass

Cable An electrical conductor surrounded by insulation. A typical cable known as twin and earth consists of two conductor wires, live and neutral and an earth wire – all normally copper or aluminium. The two conductor wires are surrounded by a plastic insulator with the bare earth wire located between them. The three wires are then covered with exterior plastic sheathing for protection. Cables are sized according to the cross-sectional area of each conductor, such as 1 mm^2 1.5 mm^2 and 2.5 mm^2.

Calorifier A cylinder or other vessel in which water is heated indirectly, normally by passing steam or hot water through a coiled pipe contained within the calorifier. An indirect hot water cylinder is a calorifier.

Cap A cover used to seal the end of a pipe run. It may fit to the pipe by means of an internal thread or alternatively by a capillary joint. *See also* Plug.

Capillary joints A fine clearance spigot and socket joint used for copper tubing in which molten solder is caused to flow by capillary attraction. *See also* Compression joint.

Cartridge fuse An electrical fuse contained in a cartridge. It is not rewireable.

Caulked joint A spigot and socket joint which is sealed. The jointing material, often cold lead forms a solid mass in the recess. Also any other sealed spigot and socket joint.

Caulking The process of forming a caulked joint.

Cavitation *See* heading under Materials and scientific principles.

Ceiling rose   An electrical lighting point attached to a ceiling. *See also* Bayonet holder.

Central heating   A heating system which provides warmth to an entire building.

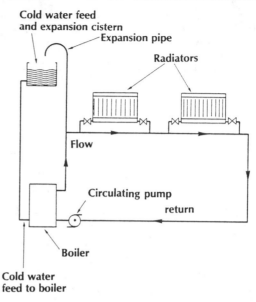

Central heating (one pipe system)

Cesspit or cesspool   An underground pit or container for the collection of foul water. It must be pumped out periodically. Normally only used in areas where mains drainage is not available. *See also* Septic tank.

Channel   An open length of pipe either semicircular or three-quarter circular in section. Mostly used in inspection chambers.

Chimney   The structure that encloses a flue.

Chlorination   A water treatment whereby chloride is added to sterilize any harmful bacteria.

Circuit   An electric cable which runs from a consumer unit or distribution board and supplies power to a number of lights, sockets or a fixed electrical appliance. Each circuit is normally protected by a circuit breaker. Also used to describe an assembly of pipes through which water flows. *See also* Ring main circuit and Radial circuit.

Circuit breaker   A device which can cut off the power to an electric circuit should a fault or overload occur. Normally either a fuse which contains a wire that melts on heating, or an electronic trip switch, known as a miniature circuit breaker (MCB).

Circulation   The flow of water through a system e.g. circulating water in a central heating system.

Cistern   An open-topped container, used for the storage of cold water at atmospheric pressure. Normally fed via a rising main and controlled with a ball valve. *See also* Indirect cold water system and Waste water preventer.

Cleaning eye   An access eye or point in waste and drainpipes. Also termed rodding eye.

Close coupled   A toilet suite where the cistern fits directly on to the pan, without any interconnecting flush pipe.

Close nipple   A short piece of pipe that is completely threaded. *See also* Nipple.

Closet   A water closet.

Coat   A single layer of a finishing material e.g. asphalt, paint, plaster and varnish etc.

Cold water service   The piped cold water supply to a building from a water main via the communication pipe, service pipe, and rising main. *See also* Direct and Indirect cold water systems.

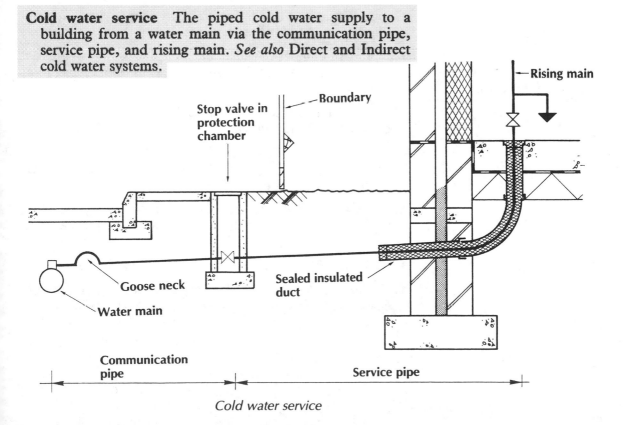

Cold water service

Combination tank A hot water storage cylinder that is combined with a cold water storage system. They are packaged together as one unit by the manufacturer. Normally used in direct cold water systems, where the small amount of stored cold water in the cistern acts as a feed for the hot water system, or for installations where space is restricted e.g. no loft space for positioning a separate cold water cistern.

Communication pipe The part of the cold water service that runs from the mains up to a stop valve in a protection chamber adjacent to the consumer's boundary. *See also* Service pipe and Gooseneck. (293)

Compression joint A screwed fitting used to joint light gauge copper, stainless steel and polythene water pipes. May be termed manipulative or non-manipulative.

Conduit Metal or plastic tubing connecting pattress boxes and through which electric cable is routed.

Conduit box A pattress box or junction box.

Consumer unit A unit combining a distribution board, circuit breakers and a mains switch. Used in domestic properties between the electric meter and the consumer's circuits.

Copper fittings Fittings for use with copper tube. *See also* Capillary and Compression joints.

Coupling or coupler A short piece of pipe having an internal or female thread used for joining the ends of screwed pipes. *See also* Nipple.

Coving A concave plaster trim fixed internally at the junction between wall and ceiling. *See also* Cornice under Architectural style.

Crazing A paint or plaster defect where the surface has cracked or split on drying.

Cross lining The horizontal hanging of a lining paper on a wall in order that the final wall covering may be hung vertically.

Curtaining The sagging of a paint film, also a concrete defect.

Cylinder A closed tank, normally cylindrical with a dome-shaped top, used for the storage of hot water prior to use. Often fitted with an immersion heater.

Daubing   Rough plastering. *See also* Dubbing out.

Dead   An electrical circuit that has been disconnected from any source of supply.

Dead leg   A length of hot water pipe in a system through which water does not circulate, except when being drawn off.

Deep seal trap   A waste pipe trap having a water seal of 75 mm or more, used to prevent induced siphonage.

Dehumidifier   A device used to reduce the moisture content or relative humidity of the air, thus reducing the likelihood of condensation. Often forms part of an air conditioning system.

Devilling   The scratching of plaster to provide a rough surface or key for the next coat.

Direct cold water system   A cold water system where all pipes to the cold water draw off points are taken direct from the service pipe. The only stored water being for a hot water system and the water closet. Only permitted in areas where there is a good supply and pressure of water. *See also* Indirect cold water system.

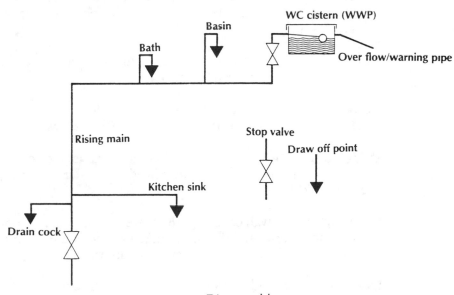

Direct cold water system

Direct hot water system   A hot water system where the cold water that flows through and is heated by the boiler or immersion heater is the water that is actually drawn off. Not suitable for use in hardwater areas as the heating process causes lime scale furring in the system. *See also* Indirect hot water system.

Distribution board   An electrical unit consisting of bus bar and circuit breakers, from which final circuits are supplied.

Distribution pipe   The pipe carrying cold water away from a storage system and distributing it to the various draw-off points.

District heating   A method of centrally heating a number of buildings from one heating source.

Double lining   The hanging of two layers of lining paper, one horizontal and the other vertical. Used to mask irregular surfaces.

Double pole   A switch in an electrical system that breaks both the live and neutral conductor.

Downpipe   A rainwater pipe (RWP).

Drain   A pipe or channel, either below or above ground, used to collect soil, waste and surface water and discharge it efficiently into a public sewer.

Drain cock or valve   A water draw-off point fitted to the lowest sections of water systems to enable complete emptying.

Drain plug   An expanding device used to temporarily seal off a portion of a drain, usually during a test.

Drain tests   A test carried out after installation to check for leakage. *See also* Air test, Smoke test and Water test.

Drip   A step formed in a flat roof at right-angles to the direction of the fall. *See also* heading under Building construction.

Drop pattern   A wall covering, the pattern of which does not repeat on a horizontal line from edge to edge of a piece but drops at each joint. *See also* Random pattern.

Dry lining   *See* heading under Building construction.

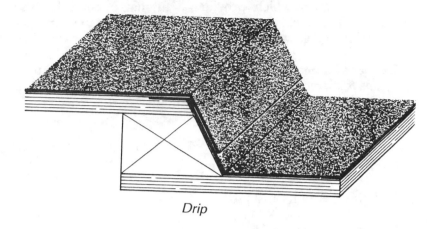

Drip

Drying out time   Period of time required for plaster to dry out before paint or wall coverings can be applied.

Dual system   A two-pipe plumbing system for drains or central heating.

Dubbing out   The application of additional material when applying the backing coat of plaster, in order to fill up hollows in an iregular background.

Earth   An earth conductor or electrical connection to the earth via an earth electrode.

Earth electrode   A conductor that is in immediate contact with, and provides an electrical connection to, the earth.

Earthing conductor   The insulated conductor which connects the earthing terminal of a distribution board or consumer unit to the earth electrode.

Eaves gutter   A rainwater gutter fixed to the eaves of a roof.

Elbow   A right-angled pipe fitting.

Expansion pipe   The vertical continuation of a hot water supply pipe to terminate over the cold water storage cistern. Also the continuation of a heating flow pipe from the boiler terminating over the cold water feed cistern. These act as ventilation pipes and provide a safe route for hot water or steam should the cylinder or boiler thermostat fail causing the water to boil. *See also* Sleeve. (292)

E

F

Expansion tank or cistern   A cistern or tank in an indirect hot water system which provides a discharge point for the expansion pipe. Normally has the dual purpose of supplying the initial cold water feed and top up to the boiler circuit.

Extract system   A mechanical ventilation system which uses fans to extract stale or contaminated air from a building. *See also* Air conditioning and Dehumidifier.

Fabric   A hessian wall covering.

Fall pipe   A rainwater pipe.

Faucet   A water tap.

Feed cistern   A cold water storage cistern which supplies cold water to the hot water system. With indirect hot water systems a combination feed and expansion cistern is used for the boiler circuit.

Ferrule   A short length of pipe or tube.

Fibrous plaster   Plaster castings and mouldings made up using plaster of Paris, reinforced with hessian and timber laths. Used for cornices, dados and other decorative ornamentation.

Field drain   Underground drainage laid with open joints or using perforated pipes. Designed to drain off surface water. Also termed a land drain.

Filler   A substance used to fill cracks and irregularities in a surface.

Fittings   In pipework any small component part (mainly joints) such as a bend, coupling, elbow and union. *See also* Capillary and Compression joints.

Flaking   The lifting or peeling of a plaster or paint coat from the coat below.

Flash   The fixing of flashings to make a weatherproof joint.

Flashing   A strip of impervious material, often sheet zinc, lead or copper, used to cover and waterproof joints in the external envelope between adjacent building elements or components such as an apron flashing, step flashing or back gutter. *See also* Flaunching and Haunching.

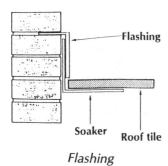

Flashing

Flat   A dried paint surface that has no gloss or sheen.

Flat-pin plug   A standard three-pin plug for use on ring main circuits, the pins of which are rectangular in section.

Flaunching   The sloping cement mortar fillet that surrounds and beds a chimney pot to its stack. *See also* Haunching.

Flex   The flexible cable which is used to connect an appliance to an electric fitting.

Float valve   A ball float valve.

Floated coat   A coat of plaster or rendering which has been finished to a smooth surface with a float. *See also* Floating.

Floating   A backing coat of plaster or rendering immediately below the finishing coat. It may be applied either direct to the background or to a render coat. Its purpose is to provide a true surface on which the finishing coat may be applied.

Flock   A wall covering with a raised pattern having a velvet-like texture.

Flow pipe   A pipe through which hot water flows from a boiler to a cylinder or heating system. *See also* Return pipe.

Flue   A pipe or passage which conveys smoke, or the products of combustion, away from a heat-producing appliance.

Flush   The process of discharging a quantity of water down a pipe, channel or sanitary appliance in order to cleanse it.

Flushing cistern   A cold water cistern that is able to flush. Used to clean water closets and urinals. Also termed a waste water preventor.

Foul water   A combination of soil water and waste water.

Foul water sewer   A sewer that conveys foul water rather than a surface water sewer. Often abbreviated to FWS.

Full-way valve   A valve that does not restrict the flow of water through the pipework when open. Used for distribution pipes and low-pressure systems.

Furred up   A blockage or restriction in plumbing systems due to the encrusted salts released from hard water when it is heated.

Fuse   An electric circuit breaker. Normally a thin piece of wire which is designed to melt when excessive current is passed through it.

Fuse box   A housing for fuses such as a consumer unit.

FWS   Foul water sewer.

G

Gas circulator   A small gas boiler that is connected to a cylinder and designed to supply it continuously with hot water.

Geyser   An instantaneous water heater which discharges hot water via a swivelling spout.

Gilding   The application of gold leaf as an ornamentation.

Gooseneck   The expansion loop in a communication pipe as it leaves the water main. (293)

Graffito   A decorative plaster surface produced by scratching a pattern in the top coat while it is still soft, to expose the lower coat which is a contrasting colour.

Graining   The painting of a surface to resemble the grain pattern of timber or the veining of marble. *See also* Rag rolling.

Grinning   A paint finish defect, where the coat below 'grins' through and is not completely obliterated. Also in plastering where joints in the background show through to the finished work, such as the mortar joints of blockwork and the abutting edges of plasterboard etc.

Ground   Earth, an electrical earth connection, and the background material or surface that is to receive a finish.

Gulley   A drainage fitting, normally trapped, over which a waste or rainwater pipe discharges; or a grated opening in a road gutter; or a hard landscaped area to receive surface water. *See also* Back inlet gulley.

Gutter   A channel used for the removal of surface water. Located at either the eaves of a roof, the edge of a hard landscaped area or the edge of a road.

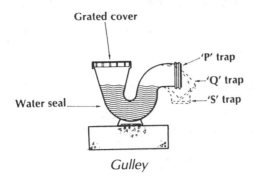

Gulley

Hard landscaping   A paved or concreted area such as a patio.

Hard water   *See* heading under Materials and scientific principles.

Haunching   The cement mortar fillet used instead of a flashing for weathering the joint between chimney stack or projecting wall and pitch roof; also the concrete surround to underground drainage pipes. *See also* Flaunching.

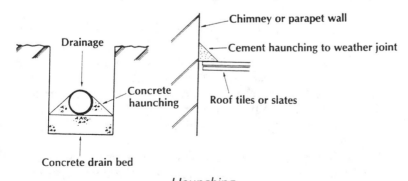

Haunching

Head   The vertical distance measured in metres from a storage cistern to a draw-off point below. This will affect the water pressure at the draw-off point.

Header tank   Normally a feed and expansion cistern of an indirect hot water system but also refers to the head tank of a sealed indirect water system for high-rise buildings which ensures a good delivery of water on the upper floors.

Holiday   A gap or miss in a coat of paint resulting from bad workmanship.

Hopper head   A hopper-shaped rainwater head.

Hungry   A surface that is too absorbent for the amount of paint or adhesive put on it. Also termed starved.

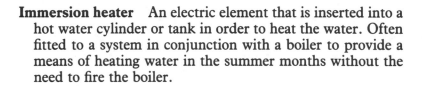

Immersion heater   An electric element that is inserted into a hot water cylinder or tank in order to heat the water. Often fitted to a system in conjunction with a boiler to provide a means of heating water in the summer months without the need to fire the boiler.

Indirect cold water system   A water system where the rising main has only one draw-off point to supply drinking water, normally at the kitchen sink. Otherwise water goes directly to the cold water storage cistern which is located in the roof space to provide sufficient head. This cistern supplies all other cold water draw-off points and a feed to the hot water system. *See also* Direct cold water system.

Indirect cylinder   The cylinder used in an indirect hot water system where the hot water from the boiler passes through a coil and does not mix with the hot water that can be drawn off. Can be termed a calorifier. *See also* Direct cylinder.

Indirect hot water system   A hot water system where the hot water from the boiler does not mix with the hot water that may be drawn off. Instead it passes through an indirect cylinder where it heats the water by conduction and convection. This is to prevent the problems of furring and corrosion caused by continually introducing fresh cold water into a boiler circuit, as happens with a direct hot water system.

Induced siphonage   The drawing or sucking away of a trap's water seal, due to the flow of water down the stack pipe. May be prevented by good design, the use of deep seal traps or anti-siphon pipes.

Inset   Basins and sinks that are fitted into a work surface, normally finishing flush.

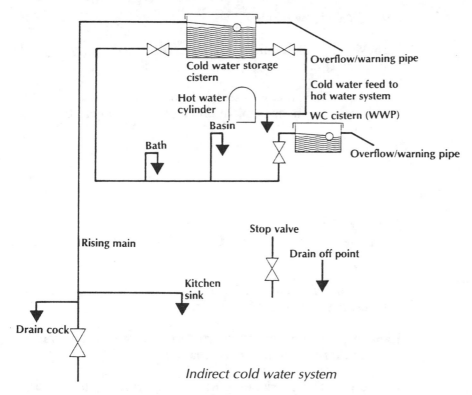

Indirect cold water system

Inspection chamber An access to underground drainage runs, provided at drain intersections and sharp bends. These allow access for inspection and the clearing of blockages. Also termed manholes.

Instantaneous water heater A water heater that heats water as it is required – it has no storage capacity. Termed either a single or multipoint heater.

Interceptor A water sealed trap which separates a drain and a sewer.

Invert The lowest level of the inside of a sewer, drain or channel at any cross-section.

Japan A black, stoved paint finish.

Joint A prepared connection between two components. May involve a fixing device as in a compression joint or a bonding agent as in a capillary joint.

Junction box A box that forms part of an electrical installation where two or more conductors join together.

Knee An elbow fitting.

Knocking up The mixing of materials. Particularly applied to mortar, plaster and fillers.

Knotting The process of sealing knots and resinous streaks in joinery and timber trim, in order to prevent them bleeding through.

Lagging The thermal insulation of water pipes, cisterns and cylinders in order to prevent freezing up or to reduce heat loss.

Land drain A drain formed using open-jointed or perforated pipes. Also termed a field drain.

Lap A paint film that slightly overlaps a previously painted surface forming a thickened coating.

Lap joint The overlapping of wall covering joints. *See also* Butt jointing.

Lath and plaster A traditional wall or ceiling finish, now obsolete, formed from timber laths, horsehair-reinforced lime mortar and a finishing coat of plaster. *See also* Lath and Lathing under Building construction.

Lavatory A place for washing, but often used to mean a washbasin or a WC pan.

Laying off The process of applying the final brush or trowel strokes to a surface that give it a smooth level finish.

Leadslate A lead flashing used to seal the joint around a pipe (often a soil stack or flue) where it penetrates through a pitched roof.

Live A conductor, circuit or electrical appliance that is connected to a source of electric power.

Live conductor The live wire. The red or brown sheathed conductor in a cable. *See also* Neutral conductor.

Lockshield valve One of the two valves attached to a radiator, it is set partially open to ensure an adequate flow of hot water.

Main   The supply authorities', gas pipe, water pipe or electric cable. Also electricity from the mains, rather than the generator or battery.

Making good   The repair of defects in a surface.

Male thread   An externally threaded fitting or pipe end that fits into an internally threaded female fitting.

Manhole   Strictly a deep inspection chamber where a person could enter for inspection and cleaning purposes.

Manipulative fitting   A compression fitting where the end of the tube has to be worked to either flare it open or form a bead. On tightening the nuts the flared end or bead will form the water seal.

Manometer   A water pressure gauge used when carrying out tests such as the air test on drainage; also called a U-tube.

Marbling   The painting of a surface to look like the veining of marble. Using a semi-transparent coat of colour over a background. *See also* Graining and Rag rolling.

Microbore or minibore   Small bore pipes of 6, 8 and 10 mm used for hot water central heating systems. *See also* Small bore.

Mixer tap   A combination tap whereby hot and cold water supplies are separately controlled but emerge from the same spout.

Mixing valve   A valve in which separate supplies of hot and cold water are mixed to a desired temperature. May be thermostatically controlled; mostly used as a control valve for showers.

Mono-block   A mixer tap having both hot and cold supplies delivered through a single pillar.

Motorized valve   An electrically-operated control valve used in hot water central heating systems.

Multi-point   An instantaneous water heater that supplies hot water to a number of taps or appliances. *See also* Single point.

Neutral conductor The neutral wire. The black or blue-sheathed conductor in a cable.

Nipple A small piece of pipe threaded at both ends. *See also* Close nipple.

Non-manipulative joint A compression joint where the seal is formed by two wedding ring-like olives which are compressed and held in place by the nuts.

Offset A double bend formed in a length of pipe so that the pipe continues in its original direction. Also termed a swan neck and seen where a rainwater pipe connects to the gutter at overhanging eaves.

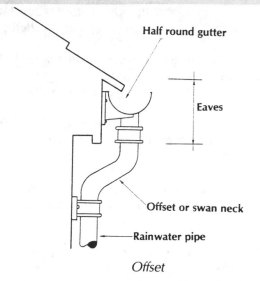

Offset

Olive A small copper or brass ring similar to a wedding ring used in non-manipulative compression fittings. On tightening the nuts of the fitting the olive compresses and forms a water seal.

One-pipe system An above-ground drainage system where both soil and waste water are discharged into one single stack. Also a hot water central heating system where the flow and returns from the radiator are connected to the same pipe. *See also* Two-pipe system.

Orange peel A paint film defect where the surface film has not flowed, causing a pock-marked or pimpled finish.

Orifice A small opening to permit the passage of a fluid.

O-ring joint   A push-fit socket and spigot joint used for waste pipes. The seal is formed by a rubber or plastic ring contained in the socket.

Outlet   A water draw-off point or an electrical socket or light fitting.

Overcloak   In sheet metal roofing the part of the upper sheet which overlaps the lower sheet or undercloak at a drip or roll joint.

Overflow pipe   *See* Warning pipe.

P

Paint   A protective and/or decorative coating applied as a liquid or plastic which later dries out or hardens to a solid film covering a surface. *See also* heading under Materials and scientific principles.

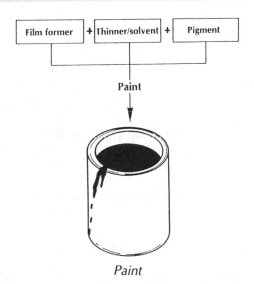

Paint

Paint system   The number of coats of paint applied to a surface including knotting, priming, undercoat and finish.

Parapet gutter   The concealed box gutter at the eaves of a roof behind a parapet wall.

Pargeting   *See* heading under Architectural style.

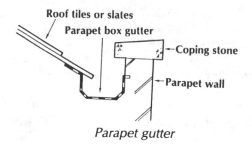

Parapet gutter

Pattern staining A discolouration of a wall or ceiling surface showing the structure behind, due to a difference in the thermal conductivity of the materials such as a ceiling where the spaces between the timber joists look dirty, caused by the fact that cold spaces between the joists conduct more heat than the joists themselves. As heat is conducted through, dust and dirt in the air is left on the surface and a striped appearance results.

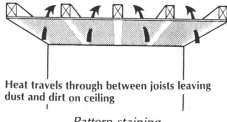

Heat travels through between joists leaving dust and dirt on ceiling

Pattern staining

Pattress The mounting box for an electrical switch or outlet. May be surface mounted or recessed to finish flush.

Pebbledash An external rendering wall finish which has had small stones thrown on its surface, shortly after application, for decorative purposes. *See also* Roughcast.

Peeling The lifting of a paint film from its backing in sheets or strips.

Pet cock Any small tap or drain cock.

Petrol intercepting chamber A form of inspection chamber used in garages. It receives surface and washing water from gulleys and separates off any oil or petrol, before allowing the water to pass into the sewerage system.

Picking up Joining live or wet edges in paint work.

Pillar tap A tap having a vertical connection as opposed to a bib tap which has a horizontal one.

Pilot light   A small gas light which is constantly lit, used to ignite the main burners of gas-fired appliances.

Plaster   The finish applied to walls and ceilings. *See also* heading under Materials and scientific principles.

Plaster base   A background to receive a plaster finish.

Plug   A small, male threaded fitting screwed into a female joint to seal it off. Also a device which enables connection between a portable electric appliance and a socket outlet. *See also* Cap.

Primary flow and return   The main pipes running to and from the boiler.

PTFE   Polytetrafluoroethylene. A plastic material in the form of either a thin tape or a paste, used as a jointing compound on pipe threads to form a seal.

Pyro   A reinforced mineral insulated cable.

Quarter bend   A ninety degree bend in a pipe, or a pipe fitting which has a ninety degree bend, such as an elbow.

Q

Radial circuit   An electric circuit for power or lighting which is connected to the consumer unit at one end only. Unlike a ring main circuit which is connected at both ends. Often used for fixed appliances which consume a lot of electricity such as cookers and immersion heaters.

R

Radiator   A container through which hot water flows. Forms part of a central heating system where it conveys heat to a room by radiation and convection. May also be fan-assisted.

Radiator key   A small socket spanner used to open the air vent when bleeding radiators of a central heating system.

Ragwork   A broken colour effect similar to graining or marbling achieved by dabbing on with a rag a semitransparent finish over a background until the desired effect is achieved.

Rainwater head A hopper or boxed-shaped enlargement at the top of a rainwater pipe used as a discharge point for other rainwater pipes or, sometimes, to collect the flow from bath and sink wastes.

Random pattern A wall hanging that has no repeat in its pattern. *See also* Drop pattern.

Reducer A boss or fitting which connects two different diameter pipes.

Render, float and set A three-coat plaster finish.

Rendering A surface coat of sand and cement mortar applied to a wall used as a waterproofing for external wall surfaces and sometimes as a backing coat for internal walls. *See also* Stucco under Architectural style.

Ring main circuit An electrical circuit for lighting and power where the two conductors link a number of lights or socket outlets together and are joined at both ends to the consumer unit. Ring mains are designed to evenly distribute the load on the circuit at any one time. Unlike a radial circuit which joins at one end only. *See also* Spur.

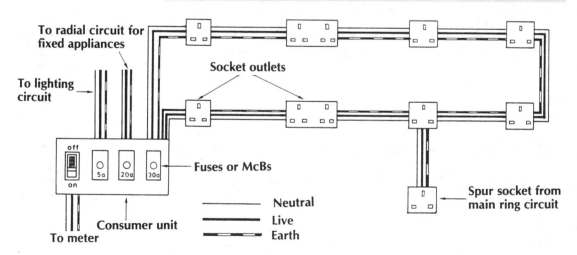

Ring main circuit

Rising main A cold water service pipe or electrical main that runs vertically through a building, supplying either water to draw-off points and cisterns, or power to consumer units or distribution boards. (293)

Rodding eye A cleaning eye to a drain run that is removed in order to pass through a set of rods when clearing a blockage.

Room-sealed appliance   An appliance having a balanced flue. It draws its air for combustion from the outside.

Ropy   A paint film which shows heavy brush marks. Also a site slang term for any poorly produced work.

Rose   The perforated head or outlet of a shower fitting. *See also* Ceiling rose.

Roughcast   An external rendering wall finish, which is roughened to produce a decorative surface. *See also* Pebble-dash.

Rubbing down   The use of an abrasive to smooth a surface prior to painting.

Running mould   A fibrous plaster moulding such as a cornice or dado that is formed in position rather than being cast and fixed later.

Runs   Dribbles of excess paint on the finished surface.

S

Saddle   A drainage fitting which sits over a large drain or sewer, making a connection.

Sanitary appliance   A fixed appliance in which water is used. Can be divided into two types – those which receive soil such as WCs and urinals and those that receive waste such as baths, sinks, washbasins and bidets.

Scratch coat   The first coat of plaster on lathing. Also a scratching coat. *See also* Lath and Plaster.

Scratching   The scoring of a plaster surface to provide a mechanical bond for the next coat. Also termed devilling. *See also* Graffito.

Screed   A line of plaster accurately laid to the finished line to provide a guide when plastering the remainder of the surface. Also a sand and cement mortar used for floor finishing.

Scrimming   The reinforcement of wall to ceiling joints and the joints of plasterboard using a woven hessian cloth set in plaster. Its purpose is to prevent cracking.

Scumbling A painting technique where the final coat is partly removed to reveal the underlying layers. Similar to ragwork.

Seal The depth of water in a trap which is designed to prevent any foul drain smells entering back into the building from a sanitary appliance.

Secret gutter A gutter which is almost completely hidden by the roofing material. Particularly applied to valley gutters.

Selvedge The edge of fabric wall covering that has to be trimmed off before hanging.

Septic tank An underground storage tank for the collection and treatment of sewage, in which the sewage slowly flows through a number of chambers where the solid content is bacterially broken down. Finally it is filtered before being discharged into a soakaway or occasionally a drainage ditch. Only used in rural area where mains sewers are not available. *See also* Cesspit.

Service pipe The portion of a water main that runs from the boundary stop valve and terminates just above the floor level in the building with another stop valve and drain down valve. (293)

Set The final coat of plaster, also termed finish.

Setting coat A finishing coat of plaster.

Sewage Foul water. *See also* Sewer.

Sewer An underground pipe or channel used for the conveyance of foul and surface water. *See also* Foul water sewer and Surface water sewer.

Sgraffito *See* Graffito.

Shading The matching of wall coverings to ensure the colouring is identical.

Shoe A fitting attached to the base of a rainwater pipe to ease the water away from the building and into an open grated gulley.

Short circuit An accidental connection between the live and neutral conductors of a circuit.

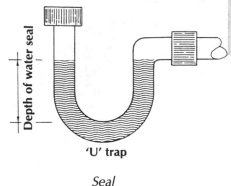

Depth of water seal

'U' trap

Seal

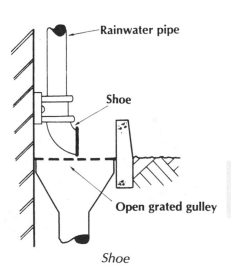

Rainwater pipe

Shoe

Open grated gulley

Shoe

Single-phase   Applied to electric circuits using a single alternating current supplied by a pair of conductors as opposed to three-phase.

Single-point heater   A geyser or instantaneous hot water heater, gas or electric powered, which supplies water to one outlet only. *See also* Multi-point heater.

Single-pole switch   A switch which breaks the live conductor wire only. As opposed to a double-pole switch.

Single stack system   An above ground drainage system where both soil and waste water are discharged into a single vertical stack pipe. *See also* Two pipe system.

Sink   A sanitary appliance used for cleaning and culinary activities. *See also* Slop sink.

Siphonic water closet   A WC which has a two-part trap. On flushing, the discharge from the cistern flows into the second stage of the trap which, as it runs away causes siphonage in the first stage of the trap, which removes most of the water along with the contents of the bowl. The trap is then resealed by the after flush. *See also* Wash down water closet.

Sizing   The application of a size to a surface which is to receive a wallcovering. This is to ensure the even absorbency of porous surfaces.

Sleeve   A pipe set through a wall or floor through which a smaller diameter pipe may be run. Also termed an expansion pipe as this method allows the smaller pipe to move without any damage or pressure being applied to the structure.

Slip   The ease with which a pasted length of wall covering moves over the surface when first put in position. The opposite of snatch.

Slopsink   A low sink, large enough to take a bucket. May also have a flushing rim similar to a WC pan. Used in hospitals for the discharge of human waste.

Small bore   A pumped, hot water central heating system utilizing mainly 15 mm pipework. *See also* Microbore.

Smoke test   A method of tracing leaks in drainage systems, normally after an air test has failed.

Snatch   The lack of movement that a length of wallcovering has when first put in position. The opposite of slip.

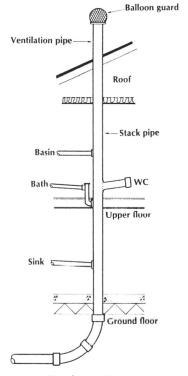

Single stack system

Soakaway A pit dug in the ground filled almost to the top with coarse hardcore and topped up flush to ground level with soil. A surface water drain is allowed to discharge in the pit, where it is able to percolate into the surrounding soil.

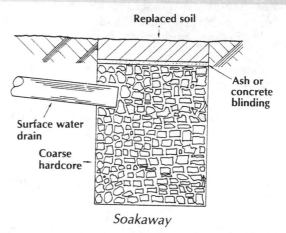

Soakaway

Soaker A small slate or tile-sized piece of sheet metal, used in conjunction with a stepped flashing to seal the joint between a pitched roof and abutting wall, such as a parapet verge wall or chimney stack.

Socket The enlarged end of a pipe into which a similar pipe spigot fits. Also another term for a coupling. *See also* Socket outlet.

Socket outlet An electric power outlet to enable connection of a portable electrical appliance by means of a plug.

Soil A term used to describe human excrement, urine and the water used to flush the system. That is the water and contents discharged from WCs and urinals.

Soil stack A drainage stack pipe.

Sparge pipe A horizontal perforated pipe used for flushing urinals.

Spatter dash A sand and cement slurry thrown on a brick or concrete background to provide a key for later plastering or rendering.

Spigot The plain end of a pipe that is inserted into a socket, as in a spigot and socket joint.

Spigot and socket joint A joint used for drainpipes, the plain spigot end of one pipe being inserted into the socket of the next.

Spraying   The application of paint or plaster using compressed air.

Spreading   The process of applying plaster with a trowel or paint with a brush.

Spur   A radial circuit that is branched off a ring main circuit. (310)

Stack   A vertical drainage or rainwater pipe. Also a chimney.

Step flashing   A flashing that covers the ends of the soakers which are interlocked with the tiles or slates where a pitched roof abuts an upstand wall. The flashing is stepped down at intervals from one mortar joint to the next into which it is dressed.

Stipple   The texturing of a paint surface by dabbing lightly with a stippling brush or plastic-covered sponge while it is still wet. Often used on ceilings.

Stop cock   A stop valve.

Stop valve   A valve either used to regulate the flow of water in a service pipe or to stop the flow completely.

Stopping   The filling or making good of joints and nail heads before painting.

Storage cistern   A cistern used for cold water storage.

Storage water heater   An appliance which heats a quantity of water and stores it for later use. Normally electrically or gas-fired. As opposed to an instantaneous heater.

Stoving   The finish of paint by the application of heat. *See also* Japan.

S-trap   A trap used for sinks, basins and WC pans, in the shape of an 'S', both the inlet and outlet being vertical.

Stripping   The removal of old paint and wallcoverings. *See also* Burning off.

Stucco   *See* heading under Architectural style.

Stuff   Normally refers to a plaster mix. As in 'knock up some stuff'.

Sullage   Waste water. *See also* Foul, Soil and Surface water.

Supply pipe   A service pipe.

Surface water   The rainwater collected from roofs and other hard landscaped areas.

Surface water sewer   A sewer that only carries surface water, sometimes termed a stormwater sewer as opposed to a foul water sewer which carries soil and waste water.

Swan neck   An offset, mainly used to describe the offset between an eave's gutter and a rainwater pipe.

Sweat   The joining of metal surfaces by allowing molten solder to flow between them, as in a capillary joint.

Sweeptee   A tee fitting where the branch curves gently away from the main run.

Switch   A mechanical device for breaking the continuity of a conductor. It turns the power off. *See also* Double pole switch and Two-way switch.

Switchboard   An assembly of switches often located adjacent to a distribution board.

SWS   A surface water sewer.

Taft joint   A joint formed between two pipes, normally lead and copper. The end of the lead pipe is opened out forming a socket into which the copper pipe is fitted. Molten solder is used to fill the gap between the two pipes sealing the joint. *See also* Wiped joint.

Tamping   The consolidation and levelling of floor screeds with a heavy straight board.

Tank   A closed rectangular storage vessel used for hot water or oil etc. Also a term used to mean a storage cistern.

Tap   The control valve used at a water draw-off point such as bib, pillar and mixer taps.

Tee   The fitting for connecting three pieces of pipe together. Takes the form of a 'T' shape. The horizontal joins the main run of pipework while the vertical, which is normally located midway along at right-angles to the horizontal, is used to connect a branch to the main run. *See also* Sweeptee.

Terminal   The end of a run such as the end of a ventilation pipe or gas flue. Also the point on an electrical fitting to which the conductor is fixed.

Thermostat   A device used as a switch to keep temperature within a set range.

Tingle   A small strip of sheet metal used to secure the edge of sheet metal covering at drips, roll and seams.

Topping   A finish applied over a screed or concrete surface, or the application of such a finish.

Trap   A means of retaining a water seal in pipework or sanitary appliances, to prevent the passage of foul drain smells into the building. Such as a U-bend, S-trap, P-trap etc.

Tumbling bay   A back drop inspection chamber often termed manhole.

Two-pipe system   An above ground drainage system consisting of two separate stack pipes, a soil stack and a waste water stack. As opposed to the single stack system which combines soil and waste water together. Also applies to central heating systems having completely separate flow and return pipes rather than a one-pipe system which uses the same pipe.

Two-way switch   A switch in a lighting circuit which enables control from two different positions. Often used either end of a flight of stairs.

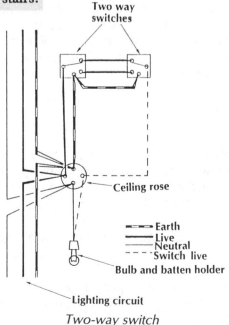

Two-way switch

Undercloak In sheet metal roofing the part of the lower sheet which is covered by the upper sheet or overcloak at a drip or roll joint. *See also* heading under Building construction.

Undercoat Any paint coat applied between the priming and finish; also any plastering coat other than the finishing coat.

Underlay The layer of felt or building paper placed between components to permit thermal movement and/or to provide a secondary weathering. Often used under sheet metal roof coverings.

Union A threaded pipe fitting consisting of conical male and female mating surfaces which can be easily connected or disconnected along the run of a pipe.

Urinal A sanitary appliance intended to receive urine only, no solids. Mainly used by men in the standing position.

Valve A device used to regulate the flow of water in a service pipe or to stop the flow completely. Also termed a stop valve, control valve or wheel valve.

Vent An outlet for air or foul smells such as a ventilation pipe.

Ventilation pipe The upper end of a stack pipe that projects above a roof in order to vent the foul drain smells. Also the upper end of a hot water service pipe which terminates over the storage cistern. The latter is normally termed an expansion pipe. (313)

Warning pipe An overflow pipe connected to all cisterns which should discharge in a conspicuous place. It provides a warning that the ball valve has failed to shut off and also a safe discharge for such water.

Wash coat A thinned or watered-down coat of emulsion paint used to seal porous unpainted surfaces.

Washdown pan A WC pan, the contents of which are emptied by a flush of water through the bowl. *See also* Siphonic pan.

Washer   A flat ring made of plastic, rubber, metal or a fibrous composition, used to form a seal or prevent slip.

Waste disposal unit   A mechanical device connected to the waste of a kitchen sink. Used for grinding kitchen garbage into a pulp allowing it to be discharged with the waste water.

Waste pipe or waste discharge pipe   A pipe that carries waste water away from an appliance.

Waste water   The discharge of water from baths, sinks, washbasins and showers. *See also* Soil, Surface and Foul water.

Waste water preventor (WWP)   A flushing cistern used to cleanse water closets and urinals.

Water closet   A sanitary appliance consisting of a pan containing a quantity of water to receive human excreta and urine, normally coupled with a flushing cistern which discharges the contents of the pan into the drainage system, cleans the surfaces and rescals the trap. Usually termed a WC or toilet.

Water softener   A device used in hard water areas to chemically remove the calcium and magnesium salts which can cause furring in hot water systems.

Water test   A test on drains which involves plugging the lower end of a drainage system, filling it up with water, then using the hydrostatic pressure to test for leaks. The water level must not fall significantly for a specified period of time.

WC   A water closet.

Weathering   *See* heading under Building construction.

Wet edge   The leading edge of a coat of paint as it is being applied to a large surface.

Wiped joint   A joint made in lead pipe work using solder where the solder is moulded around the joint with a wiping cloth. *See also* Taft joint.

WWP   A waste water preventor.

Alphabetical index